# Creativity for Engineers

**SERIES ON INDUSTRIAL AND SYSTEMS ENGINEERING**

Series Editor: Hoang Pham (*Rutgers University*)

---

*Published*

Vol. 1 Engineering Safety: Fundamentals, Techniques, and Applications
*by B. S. Dhillon*

Vol. 2 Human Reliability and Error in Medical System
*by B. S. Dhillon*

Vol. 3 Creativity for Engineers
*by B. S. Dhillon*

Series on Industrial & Systems Engineering – Vol. 3

# Creativity for Engineers

B. S. Dhillon

*University of Ottawa, Canada*

NEW JERSEY • LONDON • SINGAPORE • BEIJING • SHANGHAI • HONG KONG • TAIPEI • CHENNAI

*Published by*

World Scientific Publishing Co. Pte. Ltd.
5 Toh Tuck Link, Singapore 596224
*USA office:* 27 Warren Street, Suite 401-402, Hackensack, NJ 07601
*UK office:* 57 Shelton Street, Covent Garden, London WC2H 9HE

**British Library Cataloguing-in-Publication Data**
A catalogue record for this book is available from the British Library.

**CREATIVITY FOR ENGINEERS**



ISBN 981-256-529-9

Typeset by Stallion Press
Email: enquiries@stallionpress.com

Printed in Singapore by World Scientific Printers (S) Pte Ltd

This book is affectionately dedicated to my younger brother,
Detective Sergeant Balwant S. Dhillon.

# About the Author

Dr. B.S. Dhillon is a professor of Engineering Management in the Department of Mechanical Engineering at the University of Ottawa. He has served as a Chairman/Director of Mechanical Engineering Department/Engineering Management Program for over ten years at the same institution. He has published over 320 articles on engineering management, reliability, safety, etc. He is or has been on the editorial boards of eight international scientific journals. In addition, Dr. Dhillon has written 28 books on various aspects of engineering management, design, reliability, safety, and quality published by Wiley (1981), Van Nostrand (1982), Butterworth (1983), Marcel Dekker (1984), Pergamon (1986), etc. His books are being used in over 70 countries and many of them are translated into languages such as German, Russian, and Chinese. He has served as General Chairman of two international conferences on reliability and quality control held in Los Angeles and Paris in 1987.

Dr. Dhillon has served as a consultant to various organizations and bodies and has many years of experience in the industrial sector. At the University of Ottawa, he has been teaching engineering management (including creativity), design (including creativity), reliability, and related areas for over 26 years and he has also lectured in over 50 countries, including keynote addresses at various international scientific conferences held in North America, Europe, Asia, and Africa. In March 2004, Dr. Dhillon was a distinguished speaker at the Conference/Workshop on Surgical Errors (sponsored by White House Health and Safety Committee and Pentagon), held at the Capitol Hill (One Constitution Avenue, Washington, D.C.).

Dr. Dhillon attended the University of Wales where he received a BS in electrical and electronic engineering and an MS in mechanical engineering (management for industry). He received a Ph.D. in industrial engineering from the University of Windsor.

# Preface

Today, creativity is playing an important role than ever before in the success or failure of an organization in global competitive economy. For example, a recent survey of 500 Chief Executive Officers (CEOs) in the United States put creativity and innovation as the most important factor for survival and only 6% believed that they were in effective position to achieving it. More specifically, with regard to engineering organizations, a recent study revealed that 71% of United Kingdom (UK) engineering companies believe that creativity is very important for the success of their business.

It means that creativity is very important in engineering and all engineering professionals must possess knowledge to a certain degree in the area. Currently, in order to gain knowledge in creativity, they must study various articles, reports, or books on general creativity as there is no specific written book on creativity for engineers. To the best of the author's knowledge, there was only one book (published in 1978) that was specifically written for engineers. Needless to say, the current approach followed by engineers to gain knowledge in creativity is time consuming and difficult because of the specialized nature of the material involved.

This book is an attempt to satisfy this vital need. The material covered is treated in such a manner that the reader needs no previous knowledge to understand it. The sources of most of the material presented are given in the reference section at the end of each chapter for the benefit of the reader if he/she wishes to delve deeper into particular topics. In addition, at the end of each chapter there are numerous problems to test reader comprehension. This will allow the volume to be used as a text. A comprehensive list of references, directly or indirectly, related to creativity in engineering is provided at the end of the book, to give readers a view of the intensity of developments in the field.

The book is composed of eleven chapters. Chapter 1 presents various introductory aspects of creativity including creativity-related facts and figures, terms and definitions, creativity myths, and sources for obtaining useful information on creativity and innovation. Chapter 2 is devoted to the introductory aspects of engineering. It covers topics such as the difference between science and engineering, engineering disciplines, engineering design process, the technological team, the needs, functions, and qualities of an engineer, and the ethical and legal factors.

Chapter 3 presents famous engineering inventions, inventors, and various aspects of inventing. Creativity in organizations is presented in Chapter 4. It covers topics such as factors for the decline in corporate creativity, factors driving the need for creativity in organizations, elements of an innovative organization, creativity processes, and broad sources of information useful to creative engineers. Chapter 5 presents various different aspects of creativity management and manpower creativity including managing, selecting, and retaining creative people, tasks of key professionals in innovative companies, good creativity management qualities, creative group characteristics, and characteristics of creative and noncreative individuals.

Chapter 6 presents a total of twenty seven methods including group brainstorming, morphological analysis, synectics, checklist, force field analysis, mind mapping, and six thinking hats. Chapter 7 is devoted to creativity measurement and analysis and covers topics such as metrics for determining innovative companies' performance, fault tree analysis, control charts, cause and effect diagram, Markov method, and probability tree analysis.

Chapters 8 and 9 are devoted to creativity climate and creativity barriers, respectively. Some of the topics covered in Chapter 8 are organization's creative culture attributes, creative climate dimensions, creative work environment determinents, steps for fostering creative environment in companies, and workplace creativity climate assessment checklist. Chapter 9 includes topics such as the types of organizations finding creativity most difficult, obstacles to innovation in large organizations, management barriers to creativity, types of barriers to an individual's creative thinking, and stumbling blocks and building blocks to creativity.

Chapter 10 is devoted to creativity in many diverse areas: quality management, software development process, rail transit stations, and specific organizations. Chapter 11 presents various different aspects of creativity testing, recording, and patents. Some of the specific topics covered in the chapter are creativity tests, tools for recording ideas, and the patenting process.

This book will be useful to many people including all types of professional engineers, engineering managers, graduate and senior undergraduate students of engineering, and researchers and instructors in engineering, psychology, and business administration.

The author is deeply indebted to many individuals including colleagues, friends, and students for their invisible inputs throughout the project. I thank my children Jasmine and Mark for their patience and intermittent disturbances resulting in many desirable coffee and other breaks. Last, but not least, I thank my other half, friend, and wife, Rosy, for typing various portions of this book and other related materials, and for her timely help in proofreading.

B.S. Dhillon
Ottawa, Ontario

# Contents

Chapter 1

# Introduction

## 1.1 The Need for Creativity in Engineering

Today, creativity is playing an important role than ever before in the success or failure of an organization in global competitive economy. Thus, innovation for many companies is not an issue, but it is the only issue. A recent survey of 500 Chief Executive Officers (CEOs) in the United States put creativity and innovation as the most important factor for survival and only 6% firmly believed that they were in effective position to achieving it.[1]

More specifically, in regard to engineering organizations, a recent study revealed that 71% of the United Kingdom (UK) engineering companies believe that creativity is very important for the success of their business. The study, also, revealed that creativity and innovation is climbing higher up the agenda within the engineering sector. Some of the important factors on the need for creativity and innovation in engineering include fiercely competitive global economy, complex and sophisticated systems, alarming increase in new technologies, challenging use environments for engineering products, and short development and other times.

## 1.2 Creativity and Innovation History

Although the word "creativity" was coined only recently (i.e., the Oxford Dictionary dates its first appearance to 1875), the history of creativity can be traced back to 3500 BC when wheel first appeared in Mesopotamia and writing was invented in Sumeria.[2] In 3000 BC, Chinese invented the abacus to help individuals keep track of numbers as they do the computing and in 400 BC, Greeks invented the catapult, the first artillery weapon. In 100 AD, Chinese invented paper and in 1000 AD, they discovered a weak form of gunpowder.

In 1450 AD, Printing Press was invented by Johannes Gutenberg (1400–1468 AD) in Germany and in 1642, adding machine was devised by Blaise Pascal at the age of nineteen in France.[3,4] In the eighteenth century, items such as lightening rod, sewing machine, biofocal glasses, jigsaw puzzle, hot air balloon, and electric battery were invented.[5,6] The nineteenth century witnessed a flood of inventions including items such as dynamite, motorcycle, telephone, Kodak camera, light bulb, automobile, and flashlight. Items such as air conditioner, traffic light, airplane, microwave oven, computer mouse, Internet, and personal computer were invented in the 20th century.[3–6]

Over the centuries, many people have contributed to the field of creativity. A comprehensive list of publications for the period 1566–1974 is given in Ref. 7. These publications are categorized into eight distinct classifications: general creativity, scientific creativity, creativity of women, creativity in the fine arts, creativity and psychopathology, facilitating creativity through education, and creativity in engineering and business, industry, and development studies.

Since 1974, many publications on creativity and innovation have appeared. Many of these publications are listed at the end of this book.

## 1.3 Engineering History

The history of engineering may be traced back to the ancient times (e.g., the first appearance of wheel in Mesopotamia, the construction of Egyptians Pyramids, the construction of aqueducts and bridges by the Romans, and the construction of the Great Walls by the Chinese). In the modern times, the development of the steam engine by James Watts in 1769 may be regarded as a major milestone in the history of engineering. This invention has played an important role in Industrial Revolution. During the Industrial Revolution, the discipline of engineering developed to a stage where it started to branch out into many specialized areas.

For example, the American Institute of Mining and Metallurgical Engineers, the American Society of Mechanical Engineers, and the American Institute of Electrical Engineers were formed in 1871, 1880, and 1884, respectively. Today, there are over 22 engineering disciplines and a vast number of scientific journals and books that are playing an instrumental role in the development of the engineering field.

The detailed history of engineering is available in Refs. 8–11.

## 1.4 Creativity-Related Facts and Figures

Some of the United States, directly or indirectly, creativity-related facts and figures are as follows[12]:

- Sysco Corporation, a $23 billion food distributing company, reported that employees who took part in creativity training increased their sales between 25–30% on the average.
- APL/NOL, a major ocean shipping company, reported an impact of $46.6 million from cost reduction and avoidance, improved asset management, and revenue increase early into a creativity change program.
- Snack-food giant Frito-Lay reported more than $100 million in cost reductions due to creativity training sessions to its employees.
- Guidant, a medical equipment manufacturer, leaped into Fortune's list of the top 100 companies to work for and came thirty-first in its first try because of its creativity-related efforts.
- 3M, a diversified technology company that aggressively pursues innovation, reported that it has generated over $4 billion annually from new product introductions over the past four years.
- During the period 2000–2001, after the addition of Islands of Adventure, attendance at Universal Studies' two Florida theme parks climbed 11% and crowds at Disney's four Florida parks dropped 6%.

## 1.5 Terms and Definitions

Some of the terms and definitions, directly or indirectly, related to creativity are as follows.[13–17]

- **Creativity.** This is the mental process through which a person combines and recombines his/her previous experiences, possibly with some degree of distortion, in such a way that he/she arrives at new configurations, new patterns, and arrangements that better solve some human need.

- **Inventor.** This is an individual who generates a new idea, technology, or good and service.
- **Creative problem solving.** This is the general efforts made by any groups or individuals to think creatively to find solution to a problem.
- **Innovation.** This is a creative process in which new materials, technologies, or ideas are employed for changing either the goods and services generated/produced or the way in which they are generated/produced or distributed.
- **Creative thinking.** This is the process of generating ideas that may emphasize factors such as flexibility, originality, fluency, and elaboration in thinking.
- **Invention.** This is something invented as a product of the imagination particularly a false conception or a device, contrivance, or process originated after study and experimentation.
- **Inventive task.** This is a creative task that develops a new item/device or process by joining together previously unrelated parts in a brand new way.
- **Idea generation.** This is the production of quantities of options.
- **Lateral thinking.** This is a creative thinking process that breaks away from usual solutions and finds solutions to problems in different and unique ways.
- **Morphological thinking.** This is the form or structure of using an individual's mind for generating thought.
- **Creativity analysis.** This a language approach used for problem solving.
- **Meta thinking.** This is thinking about and creating strategies for assisting ones thinking.
- **Thinking skills.** These are learned abilities that function to facilitate thinking and are usually divided into creative or critical thinking skills.
- **Group problem solving.** This is the group process of identifying solutions to specified problems.
- **Thinking style.** This is the unconscious approach in which an individual looks at and interacts with the world.
- **Open market innovation.** This is the practice of reaching outside one's own company or organization for new product and service ideas.
- **Divergent thinking.** This is a type of thinking that breaks away from established or familiar ways of doing and seeing.

- **Group brainstorming.** This is a group method for producing many options based on factors such as the divergent thinking guidelines of differing judgment, striving for quantity, freewheeling, and building on other ideas.

## 1.6 Creativity Myths, Observations, and the Role of Innovation in Organizations

There are many myths about creativity. Some of the important ones are as follows[16]:

- **Young persons are more creative than the old persons.** In reality, as per past experiences, age is not a clear predictor of an individual's creative potential. Moreover, in the business environment, the appropriate creativity can be discovered in an individual of any age.
- **Creativity is reserved for the few flamboyant risk taking individuals.** In reality, although the willingness to take calculated risks and the ability to think in unconventional ways do play roles in creativity to a certain degree, but it does not mean that creativity is confined to high-risk endeavors, bungee jumping, etc.
- **The smarter the one is, the more creative he/she is.** In reality, intelligence and creativity correlate only to a point and once one has sufficient intelligence to do his/her job, the correlation no longer holds. More specifically, above an IQ of about 120, there is no correlation between creativity and intelligence.
- **One cannot manage creativity.** In reality, a manger can develop the environments that make creativity more likely to occur. All in all, management can make a difference!
- **Creativity is a solitary act.** In reality, as per past records, a fairly high percentage of the most important inventions are the result of collaboration among groups of individuals with complementary skills.

Four important observations about creativity are as follows[18]:

- **Difficult to practice.** It means to practice creativity calls forth, motivation, imagination, courage, confidence, inspiration, and other qualities within individuals to the near limit of their capacity.
- **Problem oriented.** It means that creativeness is manifest most clearly when an individual is faced with finding a solution to a new and difficult problem.

- **Practiced by individuals.** It means that creativity is simply a way of thinking, acting, and living peculiar to individuals.
- **Rewarding.** It means that creative work generates deeply satisfying and stimulating feelings and in turn they motivate further creative work.

Innovation in organizations plays an important role and its value in the marketplace has long been recognized as a creator and sustainer of organization/enterprise. For example, each time Intel's scientists and engineers develop a new generation of computer chips that its customers value highly, its financial fortunes are renewed. Furthermore, Intel's personal computer makes customers such as IBM, Toshiba, and Dell use the new chip to offer faster and more powerful machines to their customers. Thus, they renew their financial fortunes.

Innovation can also undermine established products and services that, in turn, can destroy enterprises. More specifically, organizations that fail to keep pace with ongoing innovations are quickly swept away from the field.

## 1.7 Useful Information on Creativity and Innovation

This section lists some of the important books, journals, conference proceedings, and organizations useful in obtaining creativity and innovation-related information.

### 1.7.1 *Books*

Some of the books that focus on creativity and innovation in engineering, directly or indirectly, are as follows:

- Bailey, R.L., *Disciplined Creativity for Engineers*, Ann Arbor Science Publishers, Inc., Ann Arbor, Michigan, 1978.
- Plsek, P.E., *Creativity, Innovation, and Quality*, ASQ Press, Milwaukee, Wisconsin, 1997.
- Addis, W., *Creativity and Innovation: The Structural Engineer's Contribution to Design*, Architectural Press, Boston, 2001.
- DeVore, H.L., *Creativity, Design, and Technology*, Davis Publications, Worcester, Massachusetts, 1989.
- Ginn, M.E., *The Creativity Challenge: Management of Innovation and Technology*, JAI Press, Greenwich, Connecticut, 1995.
- Goldenberg, J. and Mazursky, D., *Creativity in Product Innovation*, Cambridge University Press, New York, 2002.

- Hinrichs, J.R., *Creativity Scientific Research: A Critical Survey of Current Opinion, Theory, and Knowledge*, American Management Association, New York, 1961.
- Simonton, D.K., *Creativity in Science: Change, Logic, Genius, and Zeitgeist*, Cambridge University Press, New York, 2004.
- Thomas, C.A., *Creativity in Science*, Massachusetts Institute of Technology Press, Boston, 1955.
- Amabile, T.M. and Gryskiewicz, S.S., *Creativity in the R & D Laboratory, Center for Creative Leadership*, Greensboro, North Carolina, 1991.
- Benton, M.C., *Creativity in Research and Invention in the Physical Sciences: An Annotated Bibliography*, U.S. Naval Research Laboratory, Washington, D.C., 1961.


### 1.7.2 *Journals*

Some of the journals that publish articles, directly or indirectly, related to creativity and innovation are as follows:

- International Journal of Innovation and Technology Management
- Journal of Product Innovation Management
- European Journal of Innovation Management
- International Journal of Foresight and Innovation Policy
- Technovation
- Creativity Research Journal
- Journal of Creative Behaviour
- Computers and Industrial Engineering
- Automation in Construction
- Long Range Planning
- Research-Technology Management
- Computer in Industry
- Engineering Science and Educational Journal
- Artificial Intelligence
- International Journal of Technology Management
- Computer-Aided Engineering Journal
- Technological Forecasting and Social Change
- Interfaces
- Journal of Management in Engineering
- IEEE Transactions on Engineering Management
- Sloan Management Review
- Journal of Management

- Journal of Engineering and Technology Management
- IEEE Engineering Management Review
- Quality Progress
- Manufacturing Engineer
- Harvard Business Review

### 1.7.3 *Conference Proceedings*

Some of the conference proceedings that contain articles related to creativity and innovation are as follows:

- Proceedings of the IEEE International Conference on Systems, Man, and Cybernetics, 1997, 1999.
- Proceedings of the Hawaii International Conference on System Sciences, 2001.
- Proceedings of the ASME Design Engineering Technical Conference, 2002.
- Proceedings of the Third Creativity and Cognition Conference, 1999.
- Proceedings of the Frontiers in Education Conference, 2000.
- Proceedings of the Conference on Human Factors in Computing Systems, 1995.
- Proceedings of the Annual American Production and Inventory Control Society International Conference, 1997.
- Proceedings of the Creativity and Innovation Symposium, 1985.
- Proceedings of the IEEE International Symposium on Electronics and the Environment, 2000.

### 1.7.4 *Organizations*

Some of the organizations which provide information on creativity and innovation-related matters are as follows:

- American Creativity Association, P.O. Box 5856, Philadelphia, Pennsylvania 19128, USA.
- Japan Creativity Society, 27-2-105 Nando-cho, Shinjuku-ku, Tokyo, Japan.
- The European Association for Creativity and Innovation, P.O. Box 247, Enschede, 7500 AE, The Netherland.
- The International Society for Professional Innovation Management, Contact: Prof. Knut Holt, University of Trondheim, Norway (Email: Knut.Holt@iot.ntnu.no).

- Oxford Innovation Society, Isis Innovation Ltd., Ewert House, Ewert Place, Summertown, Oxford OX2 7SG, UK.

## 1.8 Scope of the Book

Today creativity and innovation has become more important than ever before as thousands of new engineering products are introduced each year in the market place and the old ones are discarded. Furthermore, a recent study conducted in the United Kingdom reported that 71% of engineering companies believe that creativity is very important for the success of their business. This basically means that creativity and innovation is a critical factor in engineering and all engineering professionals must possess knowledge to a certain degree in the area.

This book attempts to satisfy this growing need as there is no up-to-date book available in the market. The material covered in the book is up-to-date and is treated in such a manner that the reader will require no previous knowledge to understand it. Moreover, Chapters 2 and 3 provide various introductory aspects of engineering and inventions. This book will be useful to many people including all types of professional engineers, engineering managers, graduate and undergraduate students of engineering, and researchers and instructors in engineering, psychology, and business administration.

## Problems

1. Define the following terms:
   - Idea generation
   - Invention
   - Inventor
2. Write an essay on engineering history.
3. Discuss creativity and innovation history.
4. List and discuss at least five facts and figures, directly or indirectly, concerned with creativity.
5. What are the differences between creativity and innovation.
6. Discuss creativity-related myths.
7. Discuss the role of innovation in modern organizations.
8. List at least four important observations about creativity.

9. List at least five most important journals concerned with creativity and innovation.
10. Discuss the following terms:
    - Creativity analysis
    - Lateral thinking
    - Divergent thinking
    - Group brainstorming

## References

1. Colwill, J., Innovation Plus Creativity is Key, *Packaging Magazine*, Vol. 6, No. 9, 2003, p. 23.
2. Heilbron, J.L., Creativity and Big Science, *Physics Today*, November 1992, pp. 42–47.
3. Bundy, W.M., *Innovation, Creativity, and Discovery in Modern Organization*, Quorums Books, Westport, Connecticut, 2002.
4. Benton, M.C., *Creativity in Research and Invention in the Physical Sciences: An Annotated Bibliography*, U.S. Naval Research Laboratory, Washington, D.C., 1961.
5. Deutsch and Shea, Inc., *Creativity: A Comprehensive Bibliography on Creativity in Science, Engineering, Business, and the Arts*, Industrial Relations News, New York, 1958.
6. Razik, T.A., *Bibliography of Creativity Studies and Related Areas*, State University of New York at Buffalo Press, Buffalo, 1965.
7. Rothenberg, A. and Greenberg, B., *The Index of Scientific Writing on Creativity*, The Shoe String Press, Hamden, Connecticut, 1976.
8. Channell, D.F., *The History of Engineering Science: An Annotated Bibliography*, Garland Publishing, New York, 1989.
9. Hill, D.R., *A History of Engineering in Classical and Medieval Times*, Open Court Publishing Company, La Salle, Illinois, 1984.
10. Garrison, E.G., *A History of Engineering and Technology*, CRC Press, Boca Raton, Florida, 1991.
11. Fleming, A.P.M. and Brocklehurst, H.J., *A History of Engineering*, A & C Black, London, 1925.
12. Mauzy, J. and Harriman, R., *Creativity, Inc.*, Harvard Business School Press, Boston, Massachusetts, 2003.
13. Drabkin, S., Enhancing Creativity When Solving Contradictory Technical Problems, *Journal of Professional Issues in Engineering Education and Practice*, April 1996, pp. 78–82.
14. Holt, K., *Innovation: A Challenge to the Engineer*, Elsevier Science Publishers, New York, 1987.
15. Isaksen, S.G., Dorval, K.B. and Treffinger, D.J., *Creative Approaches to Problem Solving*, Kendall/Hunt Publishing Company, Dubuque, Iowa, 1994.

16. *Managing Creativity and Innovation*, Harvard Business School Publishing Corporation, Boston, Massachusetts, 2003.
17. *Merriam-Webster's Ninth New Collegiate Dictionary*, Merriam-Webster, Inc., Springfield, Massachusetts, 1989.
18. Bailey, R.L., *Disciplined Creativity for Engineers*, Ann Arbor Science Publishers, Inc., Ann Arbor, Michigan, 1978.

## Chapter 2

# Engineering: An Introduction

### 2.1 Introduction

Engineering is a critical element of our modern civilization. The term "engineering" is derived from Latin. In modern context, engineering may simply be described as the art of applying scientific and mathematical principles, experience, judgment, and common sense to make things that benefit humans.

The beginning of engineering probably occurred in Africa or Asia Minor about 8000 years ago when humans started to cultivate plants, domesticate animals, construct permanent houses in community groups, etc. The change from the nomadic life led to the need to increase food production. Thus, among the first major engineering systems were irrigation systems used for increasing crop output. The increased production in food allowed time for people to pursue other activities. For example, some became rulers, priests, and artisans. These artisans may simply be called the first engineers.[1]

One of the most remarkable early engineering accomplishments are the Egyptian pyramids. The subsequent construction of aqueducts and bridges by the Romans and the Great Walls by the Chinese are the other shining examples of the ancient engineering accomplishments.

In the modern times, the creativity and innovation of engineers played an important role in the Industrial Revolution and in the last half of the nineteenth century, they started to branch out into various specialized areas. For example, the American Institute of Mining and Metallurgical Engineers, the American Society of Mechanical Engineers, and the American Institute of Electrical Engineers were formed in 1871, 1880, and 1884, respectively.[2]

This chapter presents the various important aspects of engineering considered directly or indirectly useful to understand creativity in engineering.

## 2.2 The Difference Between Science and Engineering

The understanding of the basic distinction between science and engineering is important in order to fully appreciate the role of engineering.

Science may simply be described as a body of knowledge, particularly, man's understanding of nature and the people involved with science, i.e., scientists, direct their efforts basically to improve this understanding. They direct their efforts to find useful explanations, classifications, and means of predicting natural phenomena. More specifically, scientists spend most of their time on activities such as[2]:

- Hypothesizing explanations of natural phenomena encountered.
- Collecting data for testing the theories.
- Planning, conceiving, instrumenting, and conducting appropriate experiments.
- Performing analysis of observations and drawing effective conclusions.
- Attempting to describe the natural phenomena mathematically.
- Writing articles on the research findings.

All in all, the scientist's basic goal is knowledge as an end in itself.[2]

In contrast, the result of an engineer's efforts is usually a structure, process, or physical device. Some examples of a physical device are the weather satellite, the electronic computer, the radio telescope, and the electrocardiograph. Engineers develop such contrivances through a creative process known as design, in contrast to the scientist's central activity: research.[2]

## 2.3 Engineering Today and Engineering Disciplines

Since the time of Imhotep, Galileo, or Ampere, the basic approach to engineering has changed quite dramatically. In those times, engineering discoveries and developments were accomplished principally by individuals. Today, a team of engineers and other professionals usually work together in making engineering discoveries and developments. This is basically due to a significant increase in the population of engineers, scientists, and other professionals over the years.

For example, in the United States, the ratio of engineers and scientists to general population in 1900 was 1:1800 and in 2000, it has increased to 1:35.[1] Needless to say, engineering has played a pivotal role in modern scientific and other developments. In future, engineers must be creative and must be able to visualize what may lie ahead.

There are many engineering disciplines. In fact, the October 1983 issue of Engineering Education listed a total of 22 engineering disciplines. Nonetheless, some of the current engineering disciplines are aerospace engineering, agricultural engineering, biomedical engineering, chemical engineering, civil engineering, computer engineering, environmental engineering, electrical and electronic engineering, industrial engineering, manufacturing engineering, mechanical engineering, mining engineering, nuclear engineering, and petroleum engineering.[3] Some of these disciplines are briefly discussed below.

### 2.3.1 *Mechanical Engineering*

This is perhaps the broadest of all engineering disciplines and it applies the principles of energy and mechanics to the design of devices and machines. Nowadays, mechanical engineering professionals are involved with all types of energy conversion and utilization, machines, engines, and manufacturing materials and processes.[3]

### 2.3.2 *Civil Engineering*

This is the oldest branch of the engineering profession and it involves the use of laws, forces, and materials of nature for designing, constructing, operating, and maintaining facilities and structures such as bridges, roads, high-rise buildings, dams, sanitation plants, and water treatment centers. The specialities within the framework of civil engineering include structures, surveying, geotechnical, transportation, and sanitary and water resources.[3]

### 2.3.3 *Electrical and Electronic Engineering*

This is another important engineering discipline and is concerned with the practical application of electricity. More specifically, electrical engineers are concerned with electrical systems and devices as well as with the electrical energy. The specialty areas of electrical and electronic engineering include communications, power, electronics, measurement, and computers.

### 2.3.4 *Chemical Engineering*

This uses physics and chemistry principles to the design and production of materials that undergo chemical changes during their manufacture. Chemical engineers create, design, and operate processes that produce various types of useful materials including food products, plastics, fuels, fertilizers, fibers, health products, and structural materials.

### 2.3.5 *Aerospace Engineering*

This is concerned with developing, designing, testing, and helping to produce military and commercial aircraft, missiles and spacecraft, and new technologies in commercial aviation, space exploration, and defense systems. The specific areas of specialty include propulsion, stability and control, structures, orbital mechanics, design, testing, and aerodynamics.[3]

### 2.3.6 *Industrial Engineering*

This identifies the most effective approaches for an organization in using items such as materials, people, technology, information, energy, and processes for making or processing a product. Industrial engineering professionals plan, design, implement, and manage integrated production and service delivery systems that ensure factors such as performance, reliability, and maintainability.

### 2.3.7 *Mining Engineering*

This is another important engineering discipline and it comprises all aspects of finding, removing, and processing minerals from the earth. Mining engineering professionals design the mine layout, supervise its construction, devise appropriate systems for transporting minerals to processing plants, and develop appropriate plans for returning the concerned areas to its natural state.

### 2.3.8 *Biomedical Engineering*

This uses engineering principles and design to the biology and medical arena for improving health care and the lives of individuals with certain medical impairments. Biomedical engineers perform tasks such as design medical instruments, devices, and software; develop new procedures; perform research, and solve various types of clinical problems.

## 2.4 Engineering Design Process

This is an important element of engineering and over the years various engineering professionals have divided the design process into various steps, ranging from as few as five to as many as twenty five.[4–6] Nonetheless, the engineering design process can be categorized into the following twelve steps/stages[7]:

- Problem identification
- Problem definition
- Information gathering
- Tasks specifications
- Idea generation
- Conceptualization
- Analysis
- Experimentation
- Solution presentation
- Production
- Product distribution
- Consumption

Each of these steps is described in detail in Ref. 4.

## 2.5 The Technological Team

Modern technology has become so complex and sophisticated that it is not within the capacity of a single individual to be aware of all the intricacies of a device or process. Consequently, a technology team is formed to study and solve technological problems. The team is composed of scientists, engineers, technologists, technicians, and craftsmen. Each of these specialists is discussed separately below.

- **Scientist.** This individual searches for new knowledge concerning the nature of man and the universe. In the quest for knowledge, he/she conducts research systematically by following steps such as listed below[3]:
  - Establish a hypothesis for explaining a natural phenomenon.
  - Conduct appropriate experiments for testing the hypothesis.
  - Perform analysis of test results and state associated conclusions.

  - Generalize the hypothesis in the term of a law or theory when experimental results are considered to be in harmony with the hypothesis.
  - Publish the newly gained knowledge.
- **Engineer.** This individual is concerned with identifying and solving problems and is an innovator in applying science principles to produce cost-effective designs. In addition, he/she is often involved in supervisory work.[1]
- **Technologist.** This individual applies principles of engineering for industrial production, construction, and operation as well as works with engineering design components. The areas of interest of a technologist are typically less theoretical and mathematically-based than those of the engineering professional. Some examples of these areas are construction, production, and maintenance.
- **Technician.** This individual carries out plans and designs of engineers, performs routine equipment checks and maintenance, and sets up scientific experiments. More specifically, a technician assembles, repairs, or makes improvements in engineering equipment by gaining knowledge of its appropriate characteristics rather than by studying the scientific/engineering basis for its basic design.
- **Craftsman.** The work of this individual involves manipulative and repetitive skills requiring physical dexterity. More specifically, a craftsman possesses manual skills necessary for producing components/parts specified by engineers, scientists, technologists, and technicians. There are many different types of craftsmen including tool and die makers, carpenters, welders, precision machinists, electricians, and plumbers.[1]

## 2.6 The Needs, Functions, and Qualities of an Engineer

There are many needs of an engineer. Some of these are as follows[8–10]:

- Adequate work facilities.
- Job security.
- Adequate opportunities for self-development.
- Satisfactory work variety.
- Challenging and stimulating work.
- Adequate independence for solving a work problem.
- Proper work assignment.
- Clearly defined responsibility and authority within the organization.

- Employment with a reputable company.
- Adequate supporting staff.
- Proper recognition for his/her efforts from the management.
- Appropriate technical assistance.
- Competent bosses.
- Adequate opportunities for his/her ideas to be practiced.

Engineers perform various types of tasks or functions and they basically fall under the ten areas shown in Fig. 2.1.[3] These are design, research, construction, operations, consulting, sales, teaching, development, management, and production and testing.

In the area of design, the functions of the engineer are associated with converting a concept or model, produced by the development engineer, into a device, process, or structure. In the area of research, the functions of the engineer are associated with seeking new findings and seeking a way to use the findings. In the area of construction, the functions of the engineer are associated with various types of construction projects in the building industry. Some examples of these functions are coordinating projects, supervising projects, and estimating material, labor, and overhead costs.

In the area of operations, the functions of the engineer are associated with the facility maintenance and modifications as requirements demand. One important example of the engineer's functions in this area is to keep up with new developments in equipment so that the overhead cost is

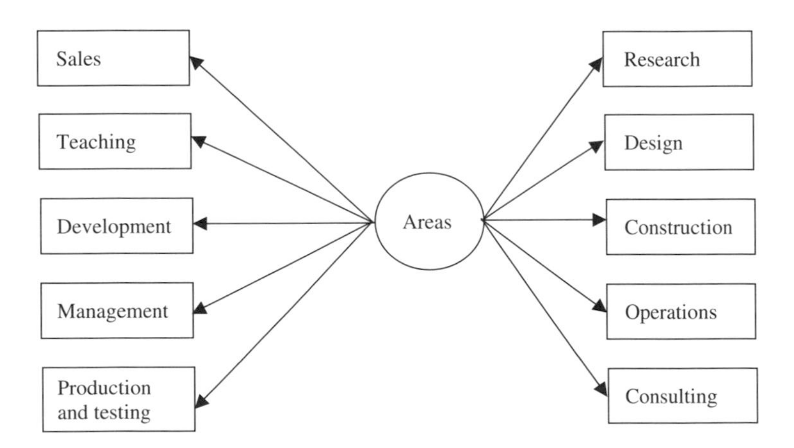


Fig. 2.1. Areas of tasks/functions performed by engineers.

maintained at the lowest possible level. In the area of consulting, the functions of the engineer are associated with providing various types of specialized help to clients upon request. A typical example of these functions is performing various types of analysis or studies on a given project.

In the area of sales, the functions of the engineer are associated with various aspects of sales. Two examples of the functions are selling one's own ideas to management that money should be allocated for the development of particular concepts or expansion of facilities and finding or creating a market for a product. In the area of teaching, the functions of the engineer are associated with imparting engineering-related knowledge to others. Usually, engineers involved with teaching have advanced degrees. Two examples of their functions are supervise research of graduate students and conduct engineering related research themselves.

In the area of development, the functions of the engineer are associated with using existing knowledge and new discoveries from research to produce a device, structure, or process that is functional. One prime example of the engineer's functions in this area is building and testing pilot models for evaluating ideas. In the area of management, the functions of the engineer are associated with items such as economics and people. Three examples of these functions are planning, coordinating, and scheduling.

In the area of production and testing, the functions of the engineer are associated with organizing production and testing facilities for the mass production of the successfully designed item. An example of the functions in this areas is to devise a schedule that will effectively coordinate materials and personnel.

An engineer must possess many qualities for his/her effective functioning. Some of these are as follows[2,9]:

- Liking for new and different things.
- Good judgment.
- Mathematical ability.
- Simulation and measurement skills.
- Skill in optimization and experimentation.
- Ability to use information sources.
- Effective communication skills.
- Ability to work with others.
- Objectivity.
- Questioning attitude.
- Capacity for self-improvement.

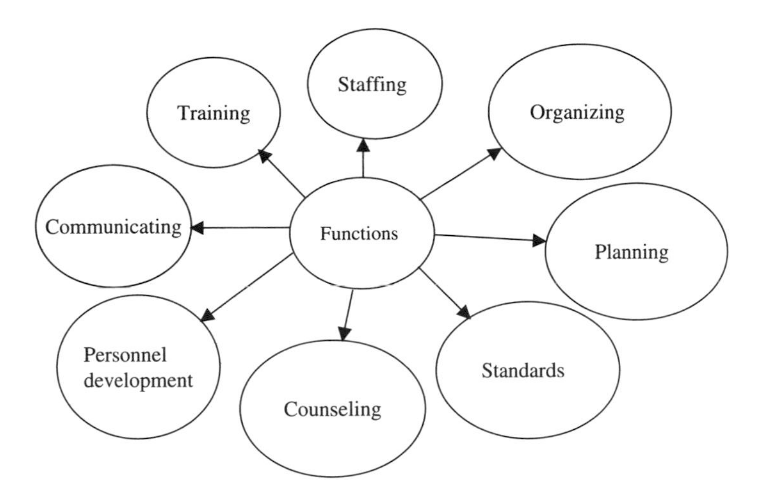


Fig. 2.2. Basic functions of an engineering manager.

## 2.7 Engineering Manager's Functions and Qualities

An engineering manager performs various functions. Some of the basic ones are shown in Fig. 2.2.[9–11] Each of the functions in Fig. 2.2 is described in detail in Ref. 11.

Some of the typical qualities of a good engineering manager are flexibility, fairness, tolerance, empathy, good humour, good emotional control, tact, self-confidence, good listening ability, ability to self-appraise, technical competence, promptness to praise and criticize, good communication skills, promptness to see good in others, ability to recognize different point of views, and freedom from suspicion and prejudice.[11]

## 2.8 The Ethical and Legal Factors

Engineering is a learned vocation and it demands high ethical standards and sound moral character of individuals practicing it.[3] It simply means that engineers must perform their tasks in an ethical manner, for winning the confidence of the public, employers, colleagues, and others. Furthermore, as engineers are involved in the use of raw materials and end products, they are probably more prone to ethical pressures than other professionals. Consequently, they are faced with questions such as: do the raw materials and end products meet ethical norms and objectives? Should they get involved

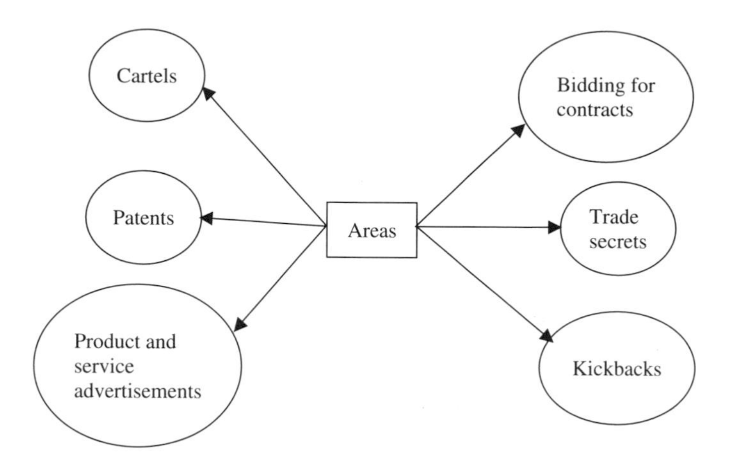


Fig. 2.3. Areas in which ethical problems may arise for engineers.

in the manufacture of products that may be harmful to humans? Some of the areas in which ethical problems may arise for engineers are shown in Fig. 2.3.[12]

Some of the typical questions with areas, as shown in Fig. 2.3, are as follows[4,12]:

- Is it quite ethical to make changes to test data results concerning a product at the insistence of your employer?
- Is it ethical to get commissions from manufacturers/contractors?
- Is it ethical to design an item that could be harmful to people?
- Is it ethical to use trade secrets of other companies to avoid bankruptcy of your company?
- Is it ethical to make use of your company's facilities for personal matters?
- Is it ethical to fix the price of goods manufactured by your company with competing companies?

To the best of author's knowledge, there is no single standard code of ethics document that can be used by all engineering professionals. This could be due to a large number of engineering societies. For example, in the United States alone, there are over 150 registered engineering or related societies.[12] Many of these societies have developed their own code of ethics to meet the need of their members.

For example, in 1912, the Institute of Electrical and Electronic Engineers (IEEE), the largest engineering professional body in the world, was the first one in the United States to develop a code of ethics for its members. The four basic articles on the IEEE code of ethics are presented in Ref. 4.

Legal factors are an important aspect of the engineering design. In recent times, changes in the interpretation of the law have resulted in an alarming increase in product liability suits. For example, in 1976, there were around 50 000 consumer-initated lawsuits in the entire United States and the prediction for the year 1980 was around 1 000 000.[13,14]

Engineering negligence accounts for 25% of product litigation and the basis for negligence in engineering design could be any of the following three factors[15]:

- Concealment of danger created by the engineering design.
- Failure of the manufacturing company to provide essential safety devices as part of the product design.
- Failure of the product design to comply with accepted norms or standards, or to specify materials of adequate strength.

Furthermore, product manufacturers in the United States may be legally liable on the basis of factors such as[5]:

- User complaints.
- Inadequate data collection with regard to failures.
- Manufacturing defects because of inadequate quality control and testing.
- Poor labeling with respect to possible danger and usage.
- Poor record keeping in regard to product manufacturing, distribution, and sale.
- Sale of the packaged, manufactured item in dangerous and incomplete form.
- Susceptibility of the instructions to detachment from the packaged, manufactured item before its sale.
- Susceptibility of the packed, manufactured item to safety-associated handling damage.

Over the years, product manufacturers have developed various points for defending against negligence charges. These points can be grouped under five distinct categories as shown in Fig. 2.4.[4] Each of these categories is described in Ref. 4.

Some of the useful guidelines that should be considered during the product design process, in order to minimize potential product liability problems, are as follows[15]:

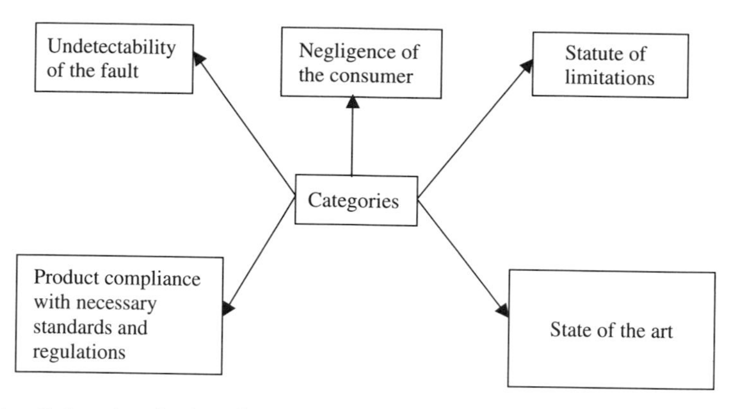


Fig. 2.4. Categories of points developed by product manufacturers for defending against negligence charges.

- Document activities such as design, manufacturing, testing, and quality control with utmost care.
- Follow applicable government and industry standards as closely as possible.
- Test all products thoroughly prior to their release for sale.
- Consider employing improved quality control methods, with the aim of lowering product liability problems.
- Make the product warning labels and the user's manual as an integral element of the product design process. Also, use international warning symbols as much as possible and involve individuals from areas such as engineering, manufacturing, legal, and marketing in developing labels and symbols.

## Problems

1. Discuss differences between science and engineering.
2. Discuss the modern day engineering discipline.
3. List at least ten distinct engineering disciplines.
4. Discuss the following engineering disciplines:
   - Biomedical engineering
   - Mechanical engineering
   - Chemical engineering
5. Discuss engineering design process.
6. List members of the technological team.

7. Discuss the needs of an engineer.
8. List at least ten important qualities of an engineer.
9. Discuss basic functions of an engineering manager.
10. Discuss areas in which ethical problems may arise for engineers.

## References


1. Beakley, G.C. and Leach, H.W., *Engineering: An Introduction to a Creative Profession*, MacMillan Publishing Company, Inc., New York, 1977.
2. Krick, E.V., *An Introduction to Engineering and Engineering Design*, John Wiley and Sons, New York, 1969.
3. Eide, A.R., Jenison, R.D., Mashaw, L.H. and Northup, L.L., *Engineering Fundamentals and Problem Solving*, McGraw-Hill Book Company, New York, 1979.
4. Dhillon, B.S., *Engineering Design: A Modern Approach*, Richard D. Irwin, Inc., Chicago, 1996.
5. Dieter, G.E., *Engineering Design*, McGraw-Hill Book Company, New York, 1983.
6. Vidosic, J.P., *Elements of Engineering Design*, The Ronald Press Company, New York, 1969.
7. Hill, P.H., *The Science of Engineering Design*, Holt, Reinhart and Winston, New York, 1970.
8. Karger, D.W. and Murdick, R.G., *Managing Engineering and Research*, Industrial Press, Inc., New York, 1969.
9. Amos, J.M. and Sarchet, B.R., *Management for Engineers*, Prentice-Hall, Inc., Englewood Cliffs, New Jersey, 1981.
10. Dhillon, B.S., *Engineering and Technology Management*, Artech House, Inc., Boston, 2002.
11. Chironis, N.P. (ed.), *Management Guide for Engineers and Technical Administrators*, McGraw-Hill Book Company, New York, 1969.
12. Walton, J.W., *Engineering Design*, West Publishing Company, New York, 1991.
13. Kolb, J. and Ross, S.S., *Product Safety and Liability: A Desk Reference*, McGraw-Hill Book Company, New York, 1980.
14. Weinstein, A.D., Twerski, A.D., Piehler, H.R. and Donaher, W.A., *Product Liability and the Reasonably Safe Products*, John Wiley and Sons, New York, 1978.
15. Dieter, G., *Engineering Design*, McGraw-Hill Book Company, New York, 1983.

## Chapter 3

# Famous Engineering Inventions, Inventors, and Inventing

### 3.1 Introduction

The term invention may simply be described as the process of devising and producing through independent investigation, mental activity, and experimentation that is useful and not previously known or exist.[1] Today, our civilization is totally surrounded by inventions and their resulting products. More specifically, our comfort, happiness, and existence are at the mercy of invention.

In fact, the ancients also highly valued inventions and discoveries. They honored their inventors by making them gods.[2] For example, the Egyptian god Osiris is said to have taught the use of plough and the art of farming, and his wife was worshipped for her discovery of barley and wheat.[2]

Although the modern inventors are not worshipped, but their inventions have become the backbone of our current civilization (e.g., steam engine (James Watt), airplane (Wright brothers), and light bulb (Thomas Edison)). This chapter presents various different aspects of inventing, including famous engineering inventions and inventors.

### 3.2 Famous Engineering Inventions

The last few centuries have witnessed a large number of engineering inventions, ranging from a pendulum clock to an airplane.[3] This section presents some of the famous engineering inventions.[3–5]

### 3.2.1 *Steam Engine*

This was designed and developed by James Watt in 1769 in Great Britain.[4] It operated on the principle of a vacuum pulling the steam down and the steam cylinder remained hot all the time.[3,5]

### 3.2.2 *Airplane*

This was designed and developed by Wilbur and Orville Wright at Kitty Hawk, North Carolina, in 1903. The Wright brothers built a 750-lb machine with a twelve horsepower motor and they made the first powered airplane in the history of mankind that lasted for twelve seconds.[6]

### 3.2.3 *Light Bulb*

This was designed and developed by Thomas Edison in 1879. In 1890, he improved the 1879 version to have all the important characteristics of a modern light bulb (i.e., an incandescent filament in an evacuated glass bulb with a screw base). These light bulbs were first used on the steamship "Columbia" and in a large New York city factory.[7]

### 3.2.4 *Radio*

This was designed and developed by Guglielmo Marconi in 1895. In 1895, Marconi sent and received signals beyond the range of vision and ultimately up to two miles. Nonetheless, it was Heinrich Hertz who discovered and first produced radio waves in 1888.[6]

### 3.2.5 *Telephone*

This was designed and developed by Alexander Graham Bell in 1876. He demonstrated it by transmitting sound clearly between Cambridge and Salem, Massachusetts.[6] The basic unit of Bell's invention was composed of a transmitter, a receiver, and a single connecting wire.

### 3.2.6 *Telescope (Reflecting)*

This was designed and constructed by Sir Isaac Newton in 1669 in Great Britain.[3]

### 3.2.7 *Motor Car*

This was designed and developed by Karl Benz in 1885 in Germany.[3]

### 3.2.8 *Telegraph*

This was invented by Samuel Morse. More specifically, he conceived an electromagnetic telegraph in 1832 and three years later, in 1835, constructed an experimental version. In 1843, the United States Congress appropriated $30 000 to Morse for constructing a telegraph line between Washington, D.C., and Baltimore, Maryland. He sent the first message: "What hath God wrought" through this line on 24 May 1844.[6]

### 3.2.9 *Alternating Current (AC) Power System*

This was devised by Nikola Tesla in 1882.[6]

### 3.2.10 *Direct Current (DC) Electric Generator*

This was designed and constructed by Michael Faraday in 1831 in Great Britain.[3]

### 3.2.11 *Typewriter*

This basic office machine was designed and constructed by Christopher Sholes in 1868 in the Untied States. The Remington Arms Company marketed it in 1873.[8]

### 3.2.12 *Sewing Machine*

This was designed and constructed by Walter Hunt in 1832 in the United States. His machine did not emulate hand sewing and it made a lock stitch using two spools of thread and incorporated an eye-pointed needle.

### 3.2.13 *Helicopter*

This flying machine was designed and developed by two bothers named Louis and Jacques Breguet in France. Their craft flew for one minute on 24 August 1907 (some sources say 29 September 1907). Nonetheless, this flight is regarded as the first vertical flight.

### 3.2.14 *Submarine*

This was designed and developed by David Bushnell in 1775/1776 in the United States. This was built for the military purpose and Bushnell called it "the Turtle". This single-passenger submarine was a tarred, oaken sphere banded with iron and powered by hand-operated pumps and propellers.[6]

### 3.2.15 *Pendulum Clock*

This was designed and constructed by Christian Huygens, an astronomer, mathematician, and physicist, in 1656 in the Netherland for measuring exact time to observe the heavens.[6]

### 3.2.16 *Zipper*

This was invented by Whitcomb L. Judson in the United States. He patented it on 29 August 1893 as a "hookless fastener". By the 1920s, zippers were used widely in clothing, luggage, and many other applications.[6]

### 3.2.17 *Bicycle (with Pedals)*

This was designed and constructed by Kirkpatrick MacMillan, a Scottish blacksmith, in 1839 in Great Britain. This two-wheeled vehicle allowed people to ride without touching their feet on the ground.[6,9–11]

### 3.2.18 *Gas Turbine*

This was designed and developed by John Barber in 1791 in Great Britain. His design contained a compressor, a combustion chamber, and a turbine. More specifically, the basics of the modern gas turbine.[9,10]

### 3.2.19 *Power Loom*

This was invented by Edmund Cartwright in 1785 in Great Britain. By 1850, there were around 250 000 cotton power-looms in Britain. In fact, it may be added that the power loom became one of the machines that played an instrumental role in the Industrial Revolution.[6]

### 3.2.20 *Hydraulic Press*

This was designed and constructed by Joseph Bramah in 1795 in Great Britain. In fact, Bramah patented a hydrostatic machine in 1785 that lead to the development of his hydraulic press ten years later.[9–11]

### 3.2.21 *Steam Turbine*

This was designed and developed by Charles A. Parsons in 1884 in Great Britain. His invention had direct application to electricity generation and marine propulsion.[9–11]

### 3.2.22 *Safety Lamp*

This was designed and developed by Humphry Davy in 1815 in Great Britain to improve the miners' safety. His invention contained no glass, the paraffin flame was encased in a fine copper mesh that lowered the surrounding air temperature to below that of the methane combustion point.[6]

### 3.2.23 *Television*

This was invented by John L. Baird in 1926 in Great Britain. His television system used a combination of electronic circuitry and a large, rotating wheel through which the picture was viewed. Thus, this television was known as "mechanical" television.[6]

### 3.2.24 *Cathode Ray Tube*

This was invented by William Crookes in 1878 in Great Britain. He used a sealed glass tube to demonstrate the path traveled by cathode rays. When he applied electric current to the tube, a patch of fluorescent light appeared on the walls of the tube due to electron interactions with residual gas in the tube.

### 3.2.25 *Hovercraft*

This was designed and developed by Cristopher Cockerell in 1956 in Great Britain. His first model was composed of a cat food can inside a coffee can in a bucket, powered by a vacuum cleaner.[9–11]

## 3.3 Famous Engineering Inventors

Over the past couple of centuries, there have been many people who invented various engineering items. This section briefly discusses some of these individuals.[5,6]

### 3.3.1 *James Watt* (1736–1819)

He was born in Scotland and worked as an instrumental maker for Glasgow University. In 1764, Watt became interested in improving steam engines, invented by Thomas Newcomen and Thomas Savery. These engines were used to pump water from mines. In 1769, he patented his first great contribution concerned with condensing the spent steam outside the cylinder.[5]

Beside the invention of modern steam engine, Watt's other inventions included the rotary engine for driving various types of machinery and the steam indicator for recording the steam pressure in the engine.[6] The electrical unit, the Watt, is named after him. James Watt died on 19 August 1819 in Heathfield.[12,13]

### 3.3.2 *Thomas Edison* (1847–1931)

He was born in Milan, Ohio, USA and at the age of twelve Edison began selling newspapers on the Grand Trunk Railway. In his spare time, he experimented with printing presses and with electrical and mechanical apparatus.[6] Edison's first important invention was a telegraphic repeating instrument that allowed automatic transmission of messages over a second line without the presence of an operator.

In 1879, Edison invented electric light bulb and in 1882, he constructed the world's first large central electric-power station, located in New York City. Altogether, Edison patented over one thousand inventions including the motion picture projector and the sound-recording device.[6,14,15]

In 1928, Edison was awarded the Congressional Gold Medal "for development and application of inventions that have revolutionized civilization in the last century". Additional information on Edison's achievements is available in Refs. 14 and 15.

### 3.3.3 *Alexander Graham Bell* (1847–1922)

He was born in Edinburgh, Scotland and received his university education at the Universities of Edinburgh and London. In 1870, Bell migrated to

Canada and the following year, to the Untied States. In 1882, he became a naturalized U.S. citizen.

In 1874, while he was working to develop a multiple telegraph, he developed the basic ideas for the telephone. Two years later, in 1876, Bell invented the telephone. The following year, he founded the Bell Telephone Company.[6]

He was one of the cofounders of the National Geographic Society and served as its president for eight years (i.e., 1896–1904). In 1883, Bell founded the Journal of Science. Additional information about Bell is available in Refs. 16 and 17.

### 3.3.4 *Wright Brothers*

The Wright brothers, Wilbur (1867–1912) and Orville (1871–1948), were born in Millville, Indiana and Dayton, Ohio, respectively. As boys, they constructed simple mechanical toys and in 1888, they built a large printing press.

In 1901, the brothers tested the effects of air pressure on more than two hundred wing surfaces and two years later, in 1903, they made the first powered airplane in the history of mankind; thus launched the world into the age of aviation.[6]

In 1908, they successfully fulfilled the requirements of a contract with the United States Army Signal Corps to develop a plane that could fly at the speed of forty miles per hour for ten minutes. In 1909, Wilbur became the president of the newly formed American Wright Company and in 1901, Orville built the first ever wind tunnel. Additional information on the Wright brothers may be found in Refs. 18 and 19.

### 3.3.5 *Samuel Morse* (1791–1872)

He was born in Charlestown, Massachusetts and educated at Yale College (Yale University). In 1825, Morse founded the National Academy of Design in New York City and served as president from 1826–1845.[6]

In 1832, Morse conceived an idea of an electromagnetic telegraph and developed an experimental version in 1835. In 1844, he built a telegraph line from Baltimore to Washington, D.C. using the $30 000 fund allocated by the United States Congress for this very purpose, in 1843.

Morse also invented a code, known as the Morse code, for use with the telegraph instrument. He received many honors and died in New York City on 2 April 1872. Additional information on Samuel Morse is available in Refs. 20 and 21.

### 3.3.6 *Guglielmo Marconi* (1874–1937)

He was born in Bologna, Italy to an Italian father and Irish mother, and educated at the University of Bologna.[6] Today, Marconi is regarded as the father of the first practical radio-signaling system.

In 1890, he became interested in wireless telegraphy and by 1895, he successfully transmitted signals to a point a few kilometers away through a directional antenna. Marconi patented his system in Great Britain and in 1897 in London, he established Marconi's Wireless Telegraph Company. Two years later, in 1899, he successfully established communication across the English Channel (i.e., between France and England). Similarly, in 1901, Marconi successfully transmitted signals across the Atlantic Ocean (i.e., between Saint John's, Canada and Poldhu, England).

In 1909, Marconi received the Nobel Prize in physics, jointly with the German physicist Karl Ferdinand Braun, for his work in wireless telegraphy. Additional information about Marconi can be found in Refs. 22 and 23.

### 3.3.7 *Nicola Tesla* (1856–1943)

He was born in Smiljan, Croatia and received his higher education at the Polytechnic School in Graz, Austria and at the University of Prague, Czech Republic. In 1884, Tesla migrated to the United States, where he later became a naturalized citizen.[6] In 1888, he devised the first ever practical alternating current (AC) power system.

Tesla patented over one hundred inventions in the United States. His other inventions included the high-frequency generators and the Tesla coil. Tesla died on 7 January 1943 in New York City and in 1956, a unit of magnetic flux density in the metric system was named after him.[6]

### 3.3.8 *Karl Benz* (1844–1929)

He was born in Karlsruhe, Germany and was credited with developing the first automobile powered by an internal-combustion engine. He built his vehicle in 1885 and patented it in 1886. The vehicle had three wheels, an electric ignition, and differential gears and was water-coded.[6]

In 1888, Benz started to advertise his vehicle, but the public refused to purchase it. However, one night his wife and two sons stole the vehicle and drove it from Mannheim to Pforzheim (approximately 65 miles). As the result of this episode, the public became fascinated by it and the Benz started to sell.

In 1893, Benz changed his car to a four-wheeled design and he died on 4 April 1929 in Mannheim, Germany. Additional information about Karl Benz is available in Ref. 24.

### 3.3.9 *Walter Hunt* (1796–1859)

He was born at Martinsburg, New York and was the first of the thirteen children born to Sherman and Rachel Hunt. In 1817, Hunt earned a degree in masonry and in 1826, he patented a flax spinning machine.

In 1833, Hunt invented the first workable sewing machine. His other inventions included a machine to make rope, a castor globe to move furniture more easily, and a coal heating stove that directed heat equally in all directions.

Hunt died in New York City on 8 June 1859 and additional information about him is available in Ref. 25.

### 3.3.10 *Christopher Sholes* (1819–1890)

He was born in Danville, Pennsylvania and after completing his schooling, he served an apprenticeship with a printer for four years. Shortly after completing his apprenticeship, Sholes moved to Wisconsin with his family. In Wisconsin, Sholes worked as a printer, journalist, and editor as well as he served two terms as a Wisconsin senator.

In 1868, Sholes invented the first practical typewriter with a keyboard layout that is familiar today. Five years later, in 1873, he sold all his typewriter rights to the Remington Arms Company. Sholes died on 17 February 1890 in Milwaukee, Wisconsin.

## 3.4 Inventing Procedure

After a careful study of the responses received from a total of 710 inventors, it was concluded that the procedure followed in inventing is basically composed of seven steps as shown in Fig. 3.1.[2]

Step 1 (i.e., observe need/difficulty) is concerned with recognizing a new, latent, or unsatisfied need or difficulty. In fact, this recognition of the need acts as a spark in imitating the entire inventive activity of the inventor.

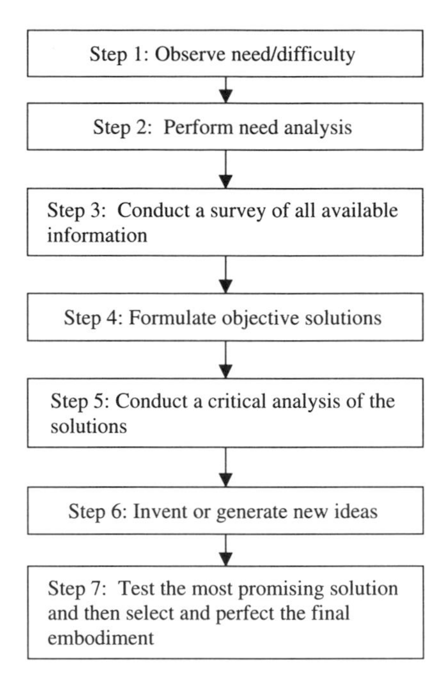


Fig. 3.1. Inventing procedure steps.

Step 2 (i.e., perform need analysis) is concerned with performing a thorough analysis of the perceived need. In fact, nearly all the surveyed 710 inventors emphasize that it is very critical not only to perceive the need but also to have a very clear idea about the object in picture.

Step 3 (i.e., conduct a survey of all available information) is concerned with surveying all available information related to the problem under consideration. This may involve a careful search through patents and the technical literature as well as a careful study of existing methods and discussions with experts, associates, producers, buyers, etc.

Step 4 (i.e., formulate objective solutions) is concerned with formulating and devising objective solutions. More specifically, all possible ways and means for meeting the identified need. These solutions include the ones that are known to the inventor or which can be developed/devised without going through the process of a wholly new conception.

Step 5 (i.e., conduct a critical analysis of the solutions) is concerned with performing a critical analysis of the objective solutions with respect to their advantages and disadvantages. It involves good reasoning, judgment, and logic.

Step 6 (i.e., invent or generate new ideas) may simply be called a mantle process that often culminates in the visualization of the solution. More specifically, it may be added that in this step, the inventor's imagination and the so-called "inventive or creative faculty" have free and full play.

Step 7 (i.e., test the most promising solutions and then select and perfect the final embodiment) is concerned with testing the tentative solution with most benefits by embodying it in physical form and determining if it really satisfies the specific needs.

## 3.5 Inventors' Characteristics, Motives, and Marital Status

The general concept of the inventor is that he/she is "different" and possesses traits that the normal individuals do not have. However, it is added that inventors differ from noninventors not on the account of any specific characteristics, but merely in the nature of their psychological reactions to shortcomings in individual's handiwork. Nonetheless, a survey on three distinct groups of individuals (i.e., patent attorneys, directors of the research and development departments, and successful inventors themselves) has identified various characteristics of inventors.[2]

More specifically, the responses (the number of responses for each characteristic are given in parentheses) of 176 patent attorneys to the question "What are the mental characteristics of inventors?", in descending order, were originality (64), analytic ability (44), imagination (34), lack of business ability (26), perseverance (20), observation (18), suspicion (12), optimism (12), and mechanical ability (6).

Similarly, the responses (the number of responses for each characteristic are given in parentheses) of the 78 directors of research and development departments in organizations, such as General Motors, Dupont, and General Electric, to the question "What are the mental characteristics of research workers and inventors?", in descending order, were analysis (48), perseverance (41), originality (37), imagination (35), training and education (20), reasoning and intelligence (20), competence (16), and observation (12).[2]

Finally, the responses (the number of responses for each characteristic are given in parentheses) of the 710 successful inventors themselves to the question "What are the characteristics of a successful inventor?", in descending order, were perseverance (503), imagination (207), knowledge and memory (183), business ability (162), originality (151), common sense

(134), analytical ability (113), self confidence (96), keen observation (61), and mechanical ability (41).[2,26]

The responses (the number of responses for each motive are given in parentheses) of 710 inventors to the question "What motives or incentives cause you to invent?", in descending order, were love of inventing (193), desire to improve (189), financial gain (167), necessity or need (118), desire to achieve (73), part of work (59), prestige (27), altruistic reasons (22), laziness (6), and no answers (33).[2]

The classifications of the martial status of the 710 inventors surveyed were married: 643, not married: 32, widower: 9, divorced: 2, and no answer: 24.[2]

## 3.6 Obstacles and Pitfalls of Inventors

There are many obstacles and pitfalls that troubled the inventors. The greatest obstacles faced by inventors are found in their external environments. Thus, they must take into consideration factors such as economic conditions, people prejudices, dishonesty of some promoters, and the problems associated with manufacturing and selling their inventions.

In descending order, the obstacles mentioned by 710 inventors (the number of their responses for each obstacle are given in parentheses) were lack of capital (136), lack of knowledge (93), prejudice (69), legal difficulties (55), marketing (54), anticipation by others (39), lack of time (38), lack of facilities (23), and no obstacles (96).[2]

There are many pitfalls that must be avoided by inventors in making their inventions. In descending order, the pitfalls mentioned by 710 inventors (the number of their responses for each pitfall are given in parentheses) were impracticability (166), overconfidence (120), lack of knowledge (112), patent attorneys (72), lack of thoroughness (46), dishonest promoters (43), discouragement (30), hope of riches (28), and disclosure to others (23).[2]

## Problems

1. What is an invention?
2. Compare the following two engineering inventions:
    - Steam engine
    - Airplane

3. Discuss the following three inventions:
   - Light bulb
   - Telescope
   - Telegraph
4. What is alternating current power system?
5. Who invented helicopter and when?
6. Who was Edmund Cartwright?
7. Discuss the following inventors:
   - Thomas Edison
   - James Watt
   - Wright brothers
8. Where was Alexander Graham Bell born?
9. Discuss the steps of the inventing procedure commonly followed by inventors.
10. Discuss the characteristics of inventors as stated by inventors themselves.

## References


1. Middendorf, W.H., *What Every Engineer Should Know About Inventing*, Marcel Dekker, Inc., New York, 1981.
2. Rossman, J., *Industrial Creativity: The Psychology of the Inventors*, University Books, Inc., New York, 1964.
3. Dhillon, B.S., *Engineering Design: A Modern Approach*, Richard D. Irwin, Inc., Chicago, 1996.
4. *The Volume Library: A Modern Authoritative Reference for Home and School Use*, The South-western Company, Nashville, Tennessee, 1993.
5. Navin, F.P.D., Engineering Creativity: Doctum Ingenium, *Canadian Journal of Civil Engineering*, Vol. 21, 1994, pp. 499–511.
6. Encarta, Microsoft Corporation, San Jose, California, 1994.
7. Cosner, S., *The Light Bulb*, Walker Publishing Company, New York, 1984.
8. Cassingham, R.C., The Dvorak Keyboard, Freelance Communications, Arcata, California, 1986.
9. *Inventions*, Time-Life Books, Alexandria, Virginia, 1994.
10. Brown, J. and Brown R., *Inventions*, Chrysalis Education, Inc., North Mankota, Minnesota, 2003.
11. Madgwick, W., *Inventions*, Kingfisher, Inc., New York, 2000.
12. Hills, R.L., *James Watt*, Landmark Publishing, Ashbourne, UK, 2002.
13. Champion, N., *James Watt*, Heinemann Library, Chicago, 2001.
14. Gomez, R., *Thomas Edison*, Abdo Publishing, Edina, Minnesota, 2003.

15. Morgan, N., *Thomas Edison*, Bookwright Press, New York, 1991.
16. Weaver, R.M., *Alexander Graham Bell*, Lucent Books, San Diego, California, 2000.
17. Pelta, K., *Alexander Graham Bell*, Silver Burdett Press, Englewood Cliffs, New Jersey, 1989.
18. Haynes, R.M., *The Wright Brothers*, Silver Burdett Press, Englewood Cliffs, New Jersey, 1991.
19. Kelly, F.C., *The Wright Brothers: A Biography*, Dover Publications, New York, 1989.
20. Hall, M.C., *Samuel Morse*, Heinemann Library, Chicago, 2004.
21. Zannos, S., *Samual Morse and the Electric Telegraph*, Mitchell Lane Publishers, Hockessin, Delware, 2004.
22. Masini, G., *Guglielmo Marconi*, Marsilio Publishers, New York, 1995.
23. Jolly, W.P., *Marconi*, Stein and Day, Inc., New York, 1972.
24. Cheney, M., *Tesla: Man Out of Time*, Prentice-Hall, Inc., Englewood Cliffs, New Jersey, 1981.
25. Kane, J.N., *Necessity's Child: The Story of Walter Hunt, America's Forgotten Inventor*, McFarland and Company, Inc., New York, 1997.
26. Bailey, R.G., *Disciplined Creativity for Engineers*, Ann Arbor Science Publishers, Inc., Ann Arbor, Michigan, 1978.

## Chapter 4

# Creativity in Organizations

### 4.1 Introduction

Creativity is an important factor in the survival of organizations in today's competitive environment. Organizations set the tone of the working environment through its organizational policies and procedures, which in turn, influence the level of creativity and productivity of their employees. A study based on interviews with 260 individuals in 22 organizations revealed that many organizations foster a play-it-safe culture with respect to creative ideas.[1] In fact, many of these organizations had better rewards for firefighting and reacting than for innovative ideas.[2]

Past experiences indicate that even when an organization puts together a team of creative individuals, the team will produce disappointing results if it is forced to function within a setup that is unfriendly to new ideas. For example, in the late 1970s and early 1980s, this situation was experienced by people in Xerox Corporation's Palo Alto Research Center (PARC).[3] The PARC scientists and engineers developed many technologies that would eventually power up desktop computers: a user-friendly operating system, the mouse, and internet connectivity. However, Xerox management overlooked these innovations because they were not going to provide financial returns within the time frame established by the company. Needless to say, many of PARC's innovations found their way into Apple personal computers. This chapter presents various important aspects of creativity in organizations.

## 4.2 Factors for the Decline in Corporate Creativity, Factors Driving the Need for Creativity in Organizations, and Organizational Creativity and Innovation Supporting Characteristics

Some of the factors for the decline in the United States Corporate creative output are as follows[4]:

- Restrictive company policies and procedures create an atmosphere that discourages "break through" attempts because of fear of failure.
- The emphasis toward narrower specialization has restricted the scope of professionals to a certain degree.
- The unpredictability of true innovative activities encourages modern master of business administration (MBA) indoctrinated management personnel to incline toward fast pay check projects at the expense of long-range growth activities.
- All-encompassing pre-inventions disclosure agreements in vogue in most American firms discourage innovative-minded individuals from pursuing new areas or concepts.
- The regulatory requirements imposed by the United States government have turned a significantly large sum of the development money away from the innovation activities.
- Corporate acquisition is frequently used for purposes such as replacing internal growth.
- The effect of the inflation factor on long-term research activities have forced many companies to incline toward fast-return projects.
- During recessionary periods, companies usually curtail their growth and innovative efforts more severely than any other area.

There are many factors that drive the need for creativity and innovation in organizations. The basic five are shown in Fig. 4.1.[2] These are association of superior long-term financial performance with innovation, increasing demand for innovation from customers, new technologies permit innovation, what used to work, does not anymore, and competitors getting better in copying past innovations.

Important organizational characteristics that support creativity and innovation are as follows[3]:

- Management accepts risk taking.
- Innovators are rewarded appropriately.

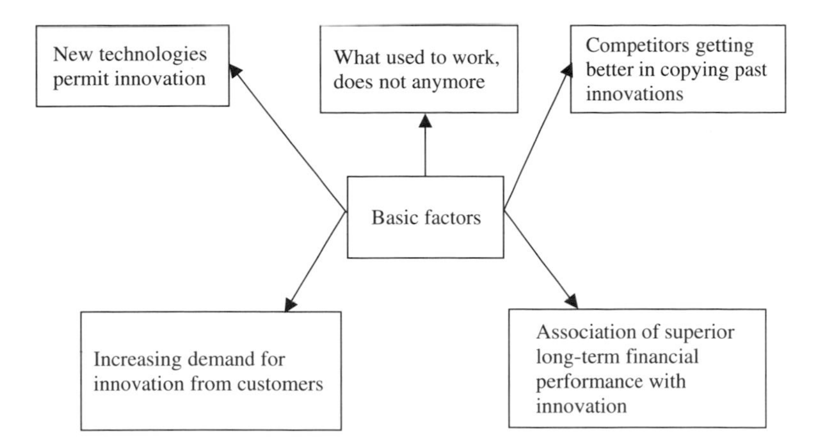


Fig. 4.1. Basic factors driving the need for creativity in organizations.

- Managers do not control the flow of information. More specifically, information flows freely.
- Employees have an effective access to knowledge sources such as customers, the scientific community, and benchmarking partners.
- Executive patrons support good ideas.
- New ideas and new ways are encouraged and welcomed.

## 4.3 Elements of an Innovative Organization

There are ten important elements of an innovative organization.[5] These are, along with their corresponding key features in parentheses, vision, leadership and the will to innovate (top management commitment, clearly articulated and shared sense of purpose, and stretching strategic intent), extensive communication (within and between the organization and external parties as well as internally in three directions: upwards, downward, and laterally), customer focus (internal and external customer orientation and total quality assurance), high involvement in innovation (participation in organization-wide continuous improvement activity), appropriate structure (organization design which allows high degrees of creativity and not always a loose "skunk works" model: key issue is given with the appropriate balance between "organic and mechanistic" options for specific contingencies), effective team working (appropriate use of teams, i.e., at local, cross-functional, and inter-organizational levels, to solve problems and needs with the investment in team selection and building), creative climate (positive approach to creative ideas, supported by appropriate reward systems), key individuals (champions, promoters, gatekeepers, and other roles which energize or facilitate innovation), learning organization (structures, processes, and cultures which help institutionalize individual learning and knowledge management), and continuing individual development (long term commitment to education and training to ensure high competence levels and the skills in an effective manner).[6]

## 4.4 Problem-Solving and Creativity Processes

Various types of problem-solving processes or methods are used by organizations for finding solutions to problems. These processes or methods must be standardized throughout the organizations in order to achieve their maximum effectiveness. Furthermore, they must be easy to understand and use by all company employees. One such problem-solving process is presented below.[7]

- Define the problem or issue effectively and avoid jumping to the cause.
- Conduct verification analysis to determine if the problem ties to the stated company/organization goals.
- Divide the problem into elements and consider reducing the scope.
- Brainstorm to identify all possible causes.
- Be prepared to collect additional data.
- Classify by major types of causes.
- Identify symptoms from root causes.
- Determine if there is a need for additional information.
- Prioritize causes by predominant cause, most probable cause, level of impact, and potential for change.
- Brainstorm for finding solutions to cause(s).
- Choose most comprehensive and easiest solution.
- Conduct cost/benefit analysis.
- Perform testing to verify the proposed solution.
- Develop an appropriate action plan (i.e., form a strong project team, assign responsibilities, develop appropriate measurements, and publish plan and progress).

Over the years, many professionals have studied the creativity process and have identified its steps from their perspective. Figure 4.2 presents the steps proposed by G. Wallas,[8] A.F. Osborn,[9] and S. Isaksen and D. Treffinger[10] in

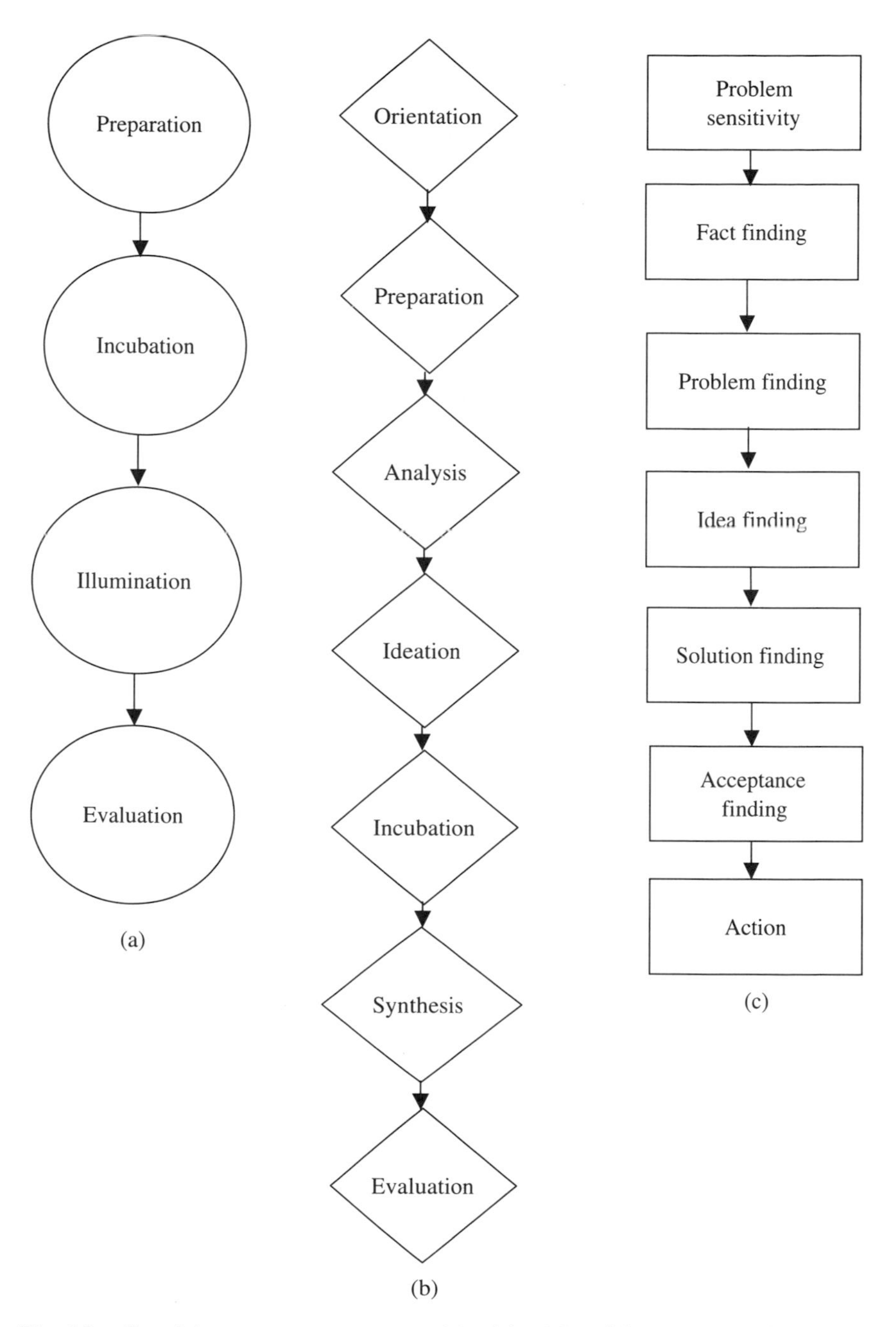


Fig. 4.2. Creativity process steps proposed by (a) Wallas, (b) Osborn, and (c) Isaksen and Treffinger.

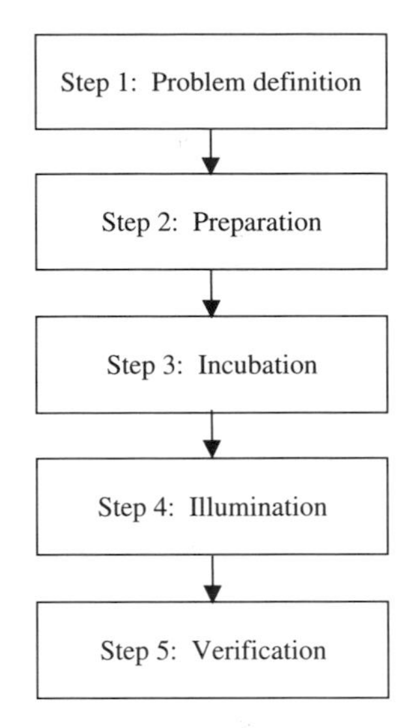


Fig. 4.3. Creativity process basic steps.

1926, 1953, and 1985, respectively.[11] However, we will discuss only the five steps of the creativity process proposed in Ref. 12 as shown in Fig. 4.3.

**Step 1:** problem definition is the recognition that there is a problem that needs to be resolved. **Step 2:** preparation is basically concerned with collecting data from various sources.[13] This information allows the creative team or individual to combine already existing ideas into new ones.[14] **Step 3:** incubation is the stage where conscious concentration on the problem under consideration ends and subconscious data processing begins. New combinations of ideas, often overlooked by noncreative teams or individuals, are examined.

**Step 4:** illumination is the moment of insight and it provides possible solutions to the problem under consideration.[15] The integrative idea may flash into the person's mind at odd times (e.g., going to sleep, waking up, or eating). Often this new idea serves as a tentative solution. Nonetheless, it may serve as a stepping stone to a more effective solution. All in all, in this step, effective communication skills are vital and failure to relay the idea properly may result in its premature death.

**Step 5:** verification is the final step of the creativity process and is concerned with a formal evaluation of the results against the proposed objective criteria.

## 4.5 Sources and Tips for Innovative Ideas and Broad Sources of Information Useful to Creative Engineers

There are many sources of innovative ideas. Some of the important ones are shown in Fig. 4.4.[3]

Customers are an evergreen source for obtaining innovative ideas if service personnel, salespeople, and research and development workers pay a careful attention to their inputs and probe for more. Needless to say, they are the best source of information when finding thc weakness of current products, identifying unsolved problems, etc.

Lead users are another important source for obtaining new ideas. They are individuals and companies (i.e., customers and non-customers) that have needs far ahead of usual market trends. Past experiences indicate that lead users are seldom interested in commercializing their innovations because they innovate simply to satisfy their own specific objectives when existing products fail to meet their needs. However, their innovations can be adopted to satisfy the needs of larger markets in future. There are various ways and means used by companies to glean innovative ideas from lead users. The following four-phase process is used by some 3M units for this very purpose[16]:

- Lay the appropriate foundation.
- Identify the trends.
- Identify lead users and learn from them.
- Develop the breakthroughs.

The process is described in detail in Ref. 16.

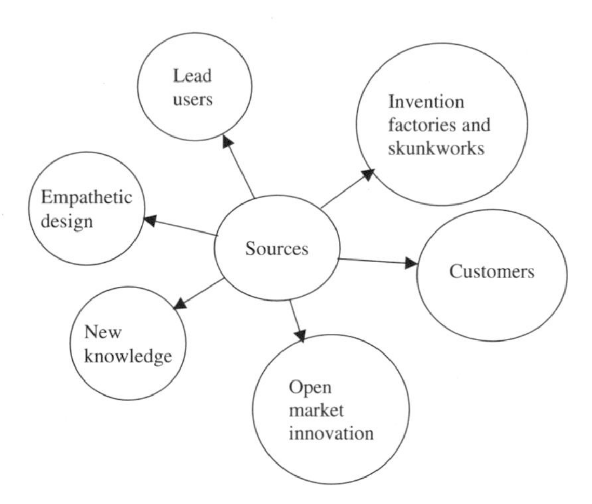


Fig. 4.4. Important sources for innovative ideas.

Past experiences indicate that many radical ideas are the result of new knowledge. Although innovations based on these ideas are frequently powerful, but there is usually a rather lengthy time span between the development of these ideas and their transformation into commercial products. All in all, despite this shortcoming, the rewards associated with new-knowledge-based innovations are often enormous.

Empathetic design is another vital source for obtaining innovative ideas. It is an ideas-generating method whereby innovators observe the usage of existing products and services by their users under their own environments. In turn, their observations become the raw material for innovative ideas. Needless to say, empathetic design is a five-step process: observe, capture data, reflect and analyze, brainstorm, and develop solution prototypes. It is described in detail in Ref. 17.

Invention factories and skunkworks are also important sources for obtaining innovative ideas. Invention factories refer to research and development units established by many large companies for generating and developing new ideas. A typical example is Bell (AT & T) Laboratories. The term skunkworks is applied to focused project teams. These teams are temporarily formed by organizations, with individuals having different perspectives, to solve a particular problem.

Open market innovation is another good source for obtaining new ideas because innovative ideas are often acquired/sold in the open market. Open market innovation employs licensing, strategic alliances, joint ventures, etc. Four important advantages of open market innovations are: importing new ideas can be useful to multiply the "building blocks" of innovation, exporting ideas provides organizations/companies a mechanism to measure an innovation's real value, exporting ideas is an effective approach for raising cash and keeping talent, and importing and exporting new ideas helps organizations/companies clarify what they do best.[18]

Besides the above six sources, some additional tips on sources for obtaining innovative ideas are shown in Fig. 4.5.[3] Additional information on these tips is available in Ref. 3.

Some of the broad sources of information useful to creative engineers are as follows[19]:

- Technical journals and magazines.
- Colleagues (i.e., conversations, private correspondences, their technical reports, etc.).
- Technical books.

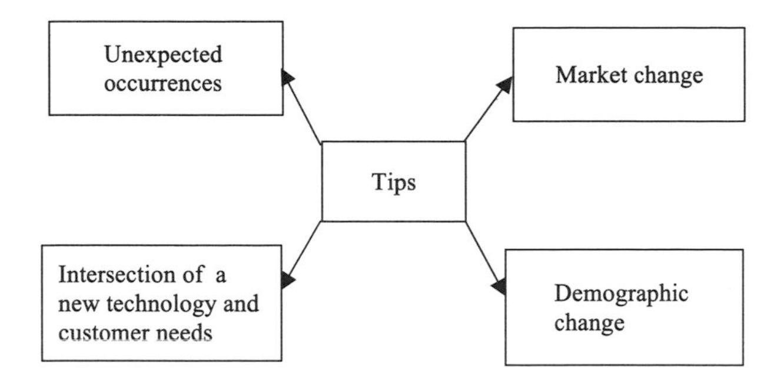


Fig. 4.5. Tips on sources for obtaining innovative ideas.

- Yourself (i.e., past experience, memory, previous observations, personal files, notebooks, publications, etc.).
- Professional societies.
- Government (i.e., technical reports, patents, etc.).
- Other companies/firms.
- Conferences/seminars.
- Management (i.e., past and current developments, confidential sources, general guidance, who is doing what and results, etc.).
- Libraries.
- Newspapers.
- Foundations and institutes.
- Clients.
- Universities.

## Problems

1. What are the important factors for the decline in corporate creativity in the United States?
2. Discuss at least five factors that are driving the need for creativity in organizations.
3. Discuss important organizational characteristics that support creativity and innovation.
4. List at least ten key elements of an innovative organization.
5. Write an essay on creativity and innovation in organizations.
6. Discuss the problem-solving process.
7. Discuss steps of a creativity process.
8. List at least six sources for obtaining new ideas.

9. Discuss at least ten broad sources of information that are useful to creative engineers.
10. Compare creativity process steps proposed by A.F. Osborn with the ones proposed by G. Wallas.

## References


1. Ryan, K.D. and Oestreich, D.K., *Driving Fear Out of the Workplace*, Jossey-Bass Publishers, San Francisco, 1991.
2. Plsek, P.E., *Creativity, Innovation, and Quality*, ASQ Quality Press, Milwaukee, Wisconsin, 1997.
3. *Harvard Business Essentials, Managing Creativity and Innovation*, Harvard Business School Publishing Corporation, Boston, Massachusetts, 2003.
4. Zeldman, M.I., How Management can Develop and Sustain a Creative Environment, in *Creativity*, edited by A. Dale Timpe, Facts on File Publications, New York, 1987, pp. 111–115.
5. Tidd, J., Bessant, J. and Pavitt, K., *Managing Innovation; Integrating Technological, Market, and Organizational Change*, John Wiley and Sons, New York, 1997.
6. Nijhof, A., Krabbendam, K. and Looise, J.C., Innovation Through Exemptions: Building Upon the Existing Creativity of Employees, *Technovation*, Vol. 22, 2002, pp. 675–683.
7. Helle, P.F., Creativity: The Key to Breakthrough Changes, How Teaming Can Harness Collective Knowledge, *Proceedings of the Annual American Production and Inventory Control Society International Conference*, 1997, pp. 301–303.
8. Wallas, G., *The Art of Thought*, Harcourt Brace and Company, New York, 1926.
9. Osborn, A.F., *Applied Imagination*, Charles Scribner's Sons, New York, 1953.
10. Isaksen, S. and Treffinger, D., *Creative Problem Solving: The Basic Course*, Bearly Limited, Buffalo, New York, 1985.
11. Navin, F.P.D., Engineering Creativity: Doctum Ingenium, *Canadian Journal of Civil Engineering*, Vol. 21, 1994, pp. 499–511.
12. Farid, F., El-Sharkawy, A.R. and Austin, L.K., Managing for Creativity and Innovation in A/E/C Organizations, *Journal of Management in Engineering*, Vol. 9, No. 4, 1993, pp. 399–409.
13. Nystrom, H., *Creativity and Innovation*, John Wiley and Sons, New York, 1979.
14. Badawy, M.K., How to Prevent Creativity Mismanagement, *IEEE Engineering Management Review*, Vol. 16, No. 2, 1988, pp. 63–70.
15. Rawlinson, J.G., *Creative Thinking and Brainstorming*, John Wiley and Sons, New York, 1981.
16. Von Hippel, E., Thomke, S. and Sonnack, M., Creating Breakthroughs at 3M, *Harvard Business Review*, September/October 1999, pp. 47–57.

17. Leonard, D. and Rayport, J.F., Spark Innovation Through Empathetic Design, Harvard Business Review, November/December 1997, pp. 102–113.
18. Rigby, D. and Zook, C., Open Market Innovation, *Harvard Business Review*, October 2002, pp. 80–89.
19. Bailey, R.L., *Disciplined Creativity for Engineers*, Ann Arbor Science Publishers, Ann Arbor, Michigan, 1978.

# Chapter 5

# Creativity Management and Manpower Creativity

## 5.1 Introduction

Just like in any other area, management plays an instrumental role in creativity. However, over the years, management in many organizations has failed to demonstrate the leadership necessary to effectively foster creativity by providing conducive environments and by personally stimulating their creative personnel.[1] For example, a study conducted by the National Science Board Committee revealed that the two important reasons for the decline in the United States technology leadership were general management practices and external financial pressures.[2,3] Needless to say, the management of many successful companies heavily rely on their creative personnel to develop new products/services that can provide a competitive edge for the company in the open market.

It simply means that manpower creativity plays a critical role in the success of an organization. The effectiveness of the manpower creativity is dictated by various factors including the characteristics of the individuals comprising the manpower, the degree of management support with respect to creativity and innovation, and the general creativity-related environment within the organization.

This chapter presents various different aspects of creativity management and manpower creativity.

## 5.2 Managing, Selecting, and Retaining Creative People

Usually, all creative people are supported by their managers or mentors who deeply encourage and influence them. The creative leadership at the top generates more trust and less jealousy among company employees. The key to the management of creativity is the management of creative individual because this individual's relationship with his/her immediate superior overshadows every other influence.[1,4]

Some of the specific suggestions for managing creative people are shown in Fig. 5.1.[4] Also, management can enhance and sustain idea generation and submission from creative people by taking actions such as follows[5]:

- Making creative people (e.g., research and development staff) aware that new project ideas are desired.
- Identifying the specific types of ideas desired.
- Communicating effectively the goals, priorities, etc.
- Making idea generation and submission an important element of the job description and performance appraisals.
- Rewarding creative people appropriately who do well at idea generation and submission.
- Exerting time pressures with respect to idea generation and submission.
- Periodically updating and communicating the information to creative people with respect to idea generation and submission.
- Publicly recognizing ideas in the event when they become a project or an element of a project.

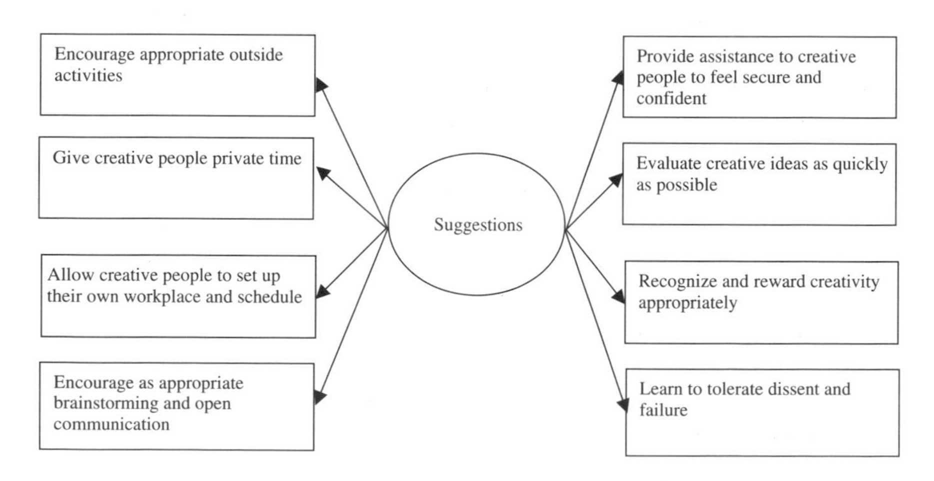


Fig. 5.1. Suggestions for managing creative people.

- Providing, as appropriate, the idea creator an opportunity for researching, developing, transferring, and implementing the project and its consequences.
- Rewarding the idea originator appropriately if the project benefits the company/organization.
- Making room, as appropriate, for the submission of incompletely developed ideas and providing necessary support to develop them fully.

In order to have a team of creative people in an organization or a unit, one must pay careful attention during job/other interviews. Some of the research-based pointers that can help differentiating the creative and less creative individuals, are as follows[6]:

- The more creative people have fewer authoritarian attitudes than the less creative individuals.
- The more creative individuals are less anxious in comparison to less creative people.
- The more creative people usually take fewer unwarranted risks than the less creative individuals.
- The more creative people are more dynamic, integrative, and autonomous in comparison to less creative individuals.
- The more creative individuals place higher values on practical matters and utility with less emphasis on elusive values, in comparison to less creative people.
- The more creative individuals are more oriented toward achievements than their less creative counterparts.
- The more creative people work slowly and cautiously when collecting data and analyzing a problem.
- The more creative people give greater evidence of psychological well-being in comparison to their less creative colleagues.
- The more creative individuals see their own attitudes as being quite different from those of other people.

As the creative people are the backbone of any organization, good managers take special measures to retain them within their establishment. Nonetheless, some of the useful guidelines for managers to retain creative individuals are as follows[1,7,8]:

- Learn to appreciate the importance of employees to the organization. Do not take them for granted.
- Publicly recognize the deserving employees by giving them appropriate credit.

- Provide employees proper opportunities for career advancement.
- Assign challenging, interesting, and meaningful tasks to employees.
- Delegate employees the authority and responsibility to perform their assigned tasks.
- Provide employees the proper opportunities for a promising future.

## 5.3 Tasks of Key Professionals in Innovative Companies

Key professionals that play an important role in innovative organizations are idea generator, champion, coach, gatekeeper, and project manager. Their tasks/responsibilities are discussed below, separately.[1,9]

### 5.3.1 *Idea Generator*

This individual performs the analysis of information about new technologies, products, or processes for generating new ideas, such as an acceptable solution to an on-going problem or the identification of a niche in the market.

### 5.3.2 *Champion*

The role of this individual is to advocate and push for new ideas by securing necessary resources and staff as well as to overcome the obstacles for positive results.

### 5.3.3 *Coach*

Some of the tasks performed by this individual are as follows:

- Handle technical and interpersonal aspects of the innovation process.
- Help all concerned individuals to work together in an effective manner for turning an idea into a tangible product/service.
- Provide appropriate technical training related to new developments.

### 5.3.4 *Gatekeeper*

This individual performs tasks such as follows:

- Collects and directs information with respect to changes in the technological environments.

- Directs relevant information to each appropriate functional area for follow-up actions.
- Uses professional conferences, the news media, personal contacts, etc., to stay abreast of new events and ideas.

### 5.3.5 *Project Manager*

This professional performs tasks such as listed below:

- Develops plans, schedules, and budgets.
- Monitors idea-related progress.
- Allocates and coordinates appropriate equipment, labor, and other resource assignments.
- Calls for appropriate periodic information sessions and status reports.

## 5.4 Good Creativity Management Qualities, Guidelines and Interventions for Managers to Improve Organizational Innovation

Good management plays an important role in building organizations where creativity becomes a way of life. Some of the qualities of good management are as follows[10]:

- It starts right from the top.
- It gets excited about ideas, concepts, developments, theories, and products.
- It appreciates mistakes.
- It builds pride.
- It does not interfere with the creative process.
- It listens.
- It encourages experimentation.
- It pays well.
- It is downright demanding.
- It encourages individuality within its demanding framework.

Some of the guidelines and interventions for managers to improve organizational innovation are as follows[11]:

- Develop innovative objectives in which all employees can have faith and they can visualize.

- Take personal responsibility for developing organizational climate for innovation and creativity.
- Constantly encourage openness and permit free play.
- Constantly seek out, encourage, and develop people with special creative talents.
- Improve frustration tolerance to errors and mistakes.
- Permit creative individuals to take as large a part as possible in the development of long term plans and in overall decision making.
- Use creativity-related performance dimensions in employee performance appraisals.
- Encourage more informal contacts between top management and idea contributors.
- Encourage a diversity of ideas and opinions.
- Dramatize problems for which creative solutions are needed.
- Recognize differences in people.
- Budget resources and time satisfactorily for innovation.
- Recognize that there is no single managerial style applicable to all individuals.
- Provide challenge by pitching projects and assignments that are just above the known capabilities of individuals.
- Allocate sufficient time for ideas to develop and mature.
- Treat mistakes and errors as opportunities for learning.
- Aim to be a resource person rather than a boss or a controller.
- Motivate individuals to come back again and again to the same problem until an effective solution is found.
- Learn to lead and enthuse by suggestions and indirect persuasion, rather than by command or specification.
- Establish specific organizational provisions for highly creative individuals and have them act as special task forces for finding solutions to complex problems.
- Teach yourself and others to respond to what is good in a proposed idea.
- Motivate individuals in calculated risk taking because it is a key element in growth and innovation.
- Occasionally permit employees to try out their pet ideas without any criticism whatsoever.
- Learn to differentiate assertiveness and aggressiveness, and act accordingly.
- Identify individuals in the organizations with a capacity to help others realizing their creative potential.

- Consider creativity as an integral part of a total organizational policy.
- Motivate employees to offer ideas concerning not only their assigned responsibilities, but also problems outside their responsibilities.
- Make changes to the seniority tradition so that promotions are possible from any level strictly on merit.
- Hold meetings and face-to-face discussions concerning opportunities that could be exploited and arrive at a commitment to particular innovative goals.
- Develop a mechanism for putting creative individuals in communication with each other, particularly across interdisciplinary lines.

## 5.5 Steps for Increasing a Manager's Own Creativity and Practical Guidelines for Creative Thinking Using Memory

Six steps for increasing a manager's own creativity are shown in Fig. 5.2.[12] Step 1: Strive for total alignment calls for the manager to align organizational goals with his/her most cherished values. Step 2: Pursue some self-initiated activity calls for the manager to select projects where his/her intrinsic motivation is high.

Step 3: Take advantage of unofficial activity as much as possible calls for nurturing idea during the unofficial period until it becomes strong enough to overcome resistance. Step 4: Try to be open to serendipity as much as possible calls for developing a bias toward action and toward trying new ideas.

Step 5: Aim to diversify your stimuli calls for intellectual cross-pollination (i.e., developing cross-functional skills and becoming a lifelong learner) for thinking in new directions. Step 6: Create opportunities for informal communication calls for taking advantage of unanticipated opportunities for exchanging ideas with colleagues.

Some of the practical guidelines for creative thinking using one's own memory are as follows[13]:

- Make use of your perception process for creating a store of ideas in memory.
- Try to come up with an original idea through novel associations of ideas you are already aware of.
- Search for different mental association by listening to other people's trains of thoughts.

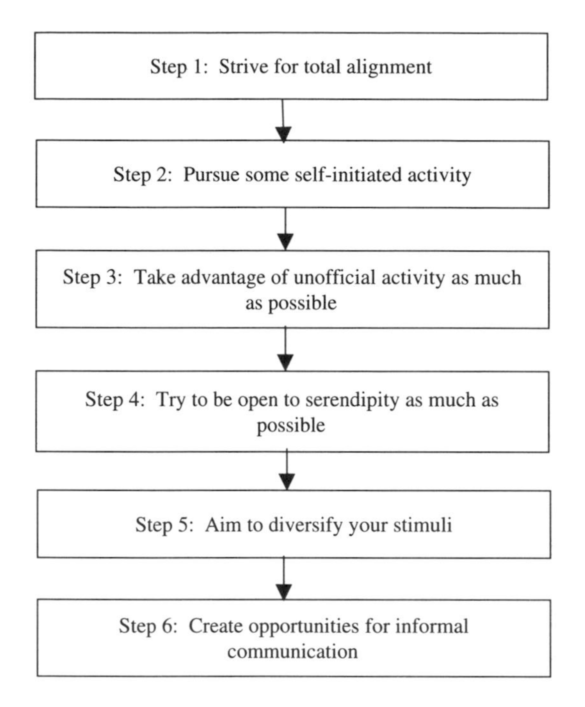


Fig. 5.2. Steps for increasing a manager's own creativity.

- Look for patterns in memory and be alert to any surprises.
- Slow down as appropriate in your thinking to highlight the intermediate ideas/concept that comprise your train of thoughts.
- Try interrupting, going slowly through, and redirecting your thoughts for determining effective or novel associations.
- Understand that your trains of thoughts are not inherently right or wrong. They are simply what you think at the moment, based primarily on your past knowledge and experience.
- Choose randomly a word and connect it to the situation of interest, and at the same time be alert to novel ideas and concepts.

## 5.6 Creative Group Characteristics and Characteristics of a Creativity Encouraging Leader

Although creativity is frequently an individual act, but past experiences indicate that many innovations are the result of creative group efforts.

Groups can frequently generate higher creative output than individuals because they bring greater insights, competencies, and vigour to the effort. Nonetheless, three important characteristics of a creative group are as follows[12]:

- **Divergent and convergent thinking.** Both divergent and convergent thinking enhance group creativity. Divergent thinking breaks away from normal or familiar ways of seeing and doing and convergent thinking stops emphasizing what is novel and instead places emphasis on what is useful.
- **Diverse thinking styles.** They enhance group creativity because group members approach their task with different preferred thinking styles.
- **Diversity of skills.** This enhances group creativity because group members bring in a range of skills.

Although the topics of characteristics of a creativity encouraging leader and good creativity management qualities (Section 5.4) may overlap here, the characteristics of a creativity encouraging leader are presented, separately, for the sake of clarity. Thus, the important characteristics of a creativity encouraging leader are shown in Fig. 5.3.[6]

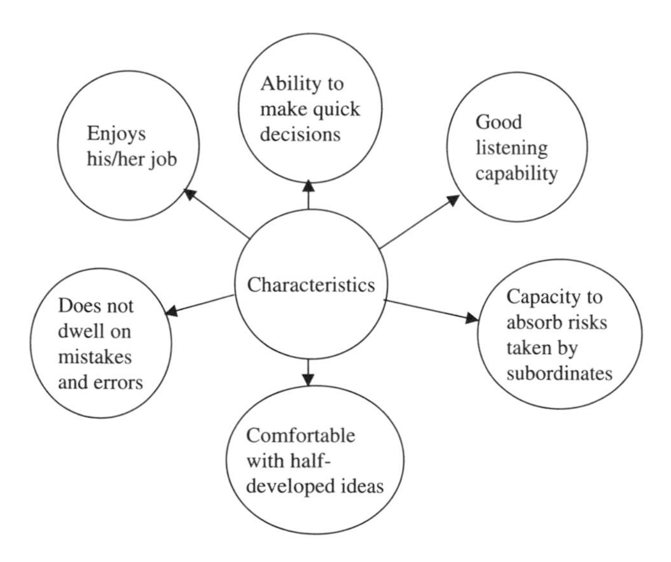


Fig. 5.3. Important characteristics of a creativity encouraging leader.

## 5.7 Characteristics of a Creative Individual, Engineer, Scientist, Entrepreneur and Characteristics of a Noncreative Person

Over the years, there has been a large number of studies concerned with determining the typical characteristics of a creative individual. As per these studies, the characteristics of a typical creative person are as follows[14–16]:

- Possesses good sense of humor.
- Possesses good listening ability.
- Works relentlessly.
- Observant.
- Nonconformist.
- Always open to experience.
- Possesses an IQ between 100 and 140.
- Takes interest in exploring ideas.
- Does not give any importance to status symbols.
- Places no value on job security.
- Accepts chaos.
- Independent.
- Likes to seek privacy and autonomy.
- Accepts failures easily.
- Insensitive to the feelings of other people.

Important characteristics of highly creative engineers are listed in Table 5.1.[16,17]

Some of the important characteristics of highly creative scientists are shown in Fig. 5.4.[18,19]

The typical characteristics of a creative entrepreneur are as follows[8]:

- He/she is neither risk-averse nor irritated by ambiguity.
- He/she needs to use his/her mind to find solutions to difficult and personally satisfying problems.
- He/she easily gets bored.
- He/she is quite comfortable with ambiguity, at least with respect to work.
- The healthier his/her personality, the more likely he/she experiences his/her work assignments as a dedicated vocation.
- He/she may not be socially "well-bounded" because of his/her lack of interest in social matters.

Table 5.1. Important characteristics of highly creative engineers.

| No. | Characteristic |
|---|---|
| 1 | Think in more abstract and theoretical terms |
| 2 | Relatively more stable |
| 3 | Possess observation and concentration power |
| 4 | Relatively less anxious |
| 5 | Question problems and new ideas |
| 6 | Have fewer close friends |
| 7 | Independent in thoughts and actions |
| 8 | Possess a high desire for freedom |
| 9 | Belong to either upper or lower 10% of engineering classes in universities or similar institutions |
| 10 | Face ambiguous situations more easily |
| 11 | Display more self-confidence |

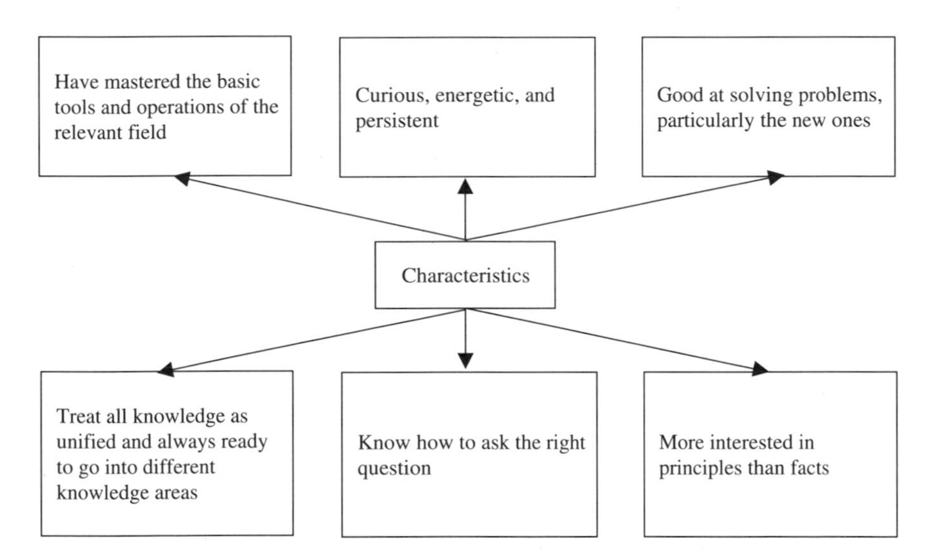


Fig. 5.4. Important characteristics of highly creative scientists.

Over the years, various creativity researchers have also studied individuals from noncreative aspects. As per their researches, the main characteristics of noncreative persons are fear of ridicule, conformity, jealous of competitiveness, cynical, resistant to change, fear of failure, seeking of security, no interest in experimenting, nonbelief in nonconventional ideas, and alienation towards systematic routine.[20,21]

## 5.8 A Successful Inventor's Characteristics' Frequency and Personal Rewards to Engineers Practicing Disciplined Creativity

A study surveyed 710 inventors to respond positively to eight characteristics of a successful inventor. Their responses to each of these characteristics in parentheses were 503 (perseverance), 207 (imagination), 183 (knowledge and memory), 162 (business ability), 151 (originality), 134 (common sense), 113 (analytic ability), 96 (self-confidence), 61 (keen observation), and 41 (mechanical ability).[22]

Some of the personal rewards to engineers practicing disciplined creativity are as follows[22]:

- Joy of accomplishment and satisfaction.
- Interesting and challenging activity as they create the new things of tomorrow's world.
- Contributing in tangible ways for human benefits.
- Greater possibility for recognition and honors.
- Greater possibility for financial rewards.
- Understanding of themselves as humans (i.e., a valuable by-product of finding solution to significant engineering problems).

## Problems

1. List and discuss at least eight useful guidelines for managing creative people.
2. Discuss the responsibilities of the following professionals in innovative organizations:
    - Idea generator
    - Gatekeeper
    - Coach
3. List at least ten qualities of good creativity management.
4. List ten most important guidelines for managers to improve organizational innovation.

5. Discuss useful steps for increasing a manager's own creativity.
6. List at least six useful practical guidelines for creative thinking using one's own memory.
7. Discuss characteristics of a creative group.
8. List at least six important characteristics of a creativity encouraging leader.
9. List at least eleven important characteristics of a creative person.
10. Discuss typical characteristics of noncreative individuals.

## References


1. Farid, F., El-Sharkawy, A.R. and Austin, L.K., Managing for Creativity and Innovation in A/E/C Organizations, *Journal of Management in Engineering*, Vol. 9, No. 4, 1993, pp. 399–409.
2. National Science Board Committee, Why U.S. Technology Leadership is Eroding, *Research-Technology Management*, January/February 1991, pp. 36–42.
3. Bundy, W.M., *Innovation, Creativity, and Discovery in Modern Organizations*, Quorum Books, Westport, Connecticut, 2002.
4. Feinberg, M.R., Fourteen Suggestions for Managing Scientific Creativity, *Research Management*, Vol. 11, No. 2, 1968, pp. 83–93.
5. Baker, N.R., Green, S.G. and Bean, A.S., How Management Can Influence the Generation of Ideas, in *Creativity*, edited by Dale Timpe, A., Facts on File Publications, New York, 1987, pp. 57–68.
6. Badawy, M.K., How to Prevent Creativity Mismanagement, in *Creativity*, edited by Dale Timpe, A., Facts on File Publications, New York, 1987, pp. 176–188.
7. Pollock, T., A Personal File of Stimulating Ideas and Problem Solvers, *Supervision*, Vol. 53, No. 5, 1992, pp. 24–26.
8. Sinetar, M., Entrepreneurs, Chaos, and Creativity: Can Creative People Really Survive Large Company Structure, in *Creativity*, edited by Dale Timpe, A., Facts on File Publications, New York, 1987, pp. 103–110.
9. Frohman, M. and Pascarella, P., Achieving Purpose-Driven Innovation, *Industry Week*, Vol. 239, No. 6, 1990, pp. 20–26.
10. Vicere, A.A., in *Creativity*, edited by Dale Timpe, A., Facts on File Publications, New York, 1987, pp. 152–155.
11. Raudsepp, E., 100 Ways to Spark Your Employee's Creative Potential, in *Creativity*, edited by Dale Timpe, A., Facts on File Publications, New York, 1987, pp. 235–243.
12. *Harvard Business Essentials: Managing Creativity and Innovation*, Harvard Business School Publishing Corporation, Boston, 2003.
13. Plsek, P.E., *Creativity, Innovation, and Quality*, ASQ Quality Press, Milwaukee, Wisconsin, 1997.

14. McPherson, J.H., Are You Creative?, *Prod. Eng.*, November 1958, pp. 28–29.
15. McPherson, J.H., The Relationship of the Individual to the Creative Process in the Management Environment, Paper No. 64MD12, American Society of Mechanical Engineers, New York, 1964.
16. Dhillon, B.S., *Engineering and Technology Management Tools and Applications*, Artech House, Inc., Boston, Massachusetts, 2002.
17. Karger, D.W. and Murdick, R.G., *Managing: Engineering and Research*, Industrial Press, Inc., New York, 1969.
18. Tang, H.K., An Integrative Model of Innovation in Organizations, *Technovation*, Vol. 18, No. 5, 1998, pp. 297–309.
19. Root-Bernstein, R.S., Who Discovers and Invents, *Research-Technology Management*, Vol. 32, 1989, pp. 43–50.
20. Harrisberger, L., *Engineermanship: A Philosophy of Design*, Wadsworth Publishing Company, Belmont, California, 1966.
21. Dhillon, B.S., *Engineering Management*, Technomic Publishing Company, Lancaster, Pennsylvania, 1987.
22. Bailey, R.L., *Disciplined Creativity for Engineers*, Ann Arbor Science Publishers, Inc., Ann Arbor, Michigan, 1978.

## Chapter 6

# Creativity Methods

### 6.1 Introduction

In today's rapidly evolving global economy, engineers and others encounter various types of problems each day. They try to overcome these problems through various ways and means including the use of creativity methods. There are a large number of creative problem-solving methods mostly developed in the 1960's and 1970's.[1] These methods are applicable to a variety of problem-solving situations.[2,3] In the early 1980's, Ref. 4 identified fifty of these methods and in 1994, Ref. 5 presented a total of one hundred and one creative problem-solving techniques.

Needless to say, there are currently a large number of creative idea generation methods in use world-wide and they can be classified under six categories: brainstorming methods, methods of creative orientation, systematic structuring, creative confrontation, systematic problem specification, and brainswriting methods.[6] This chapter presents many of these methods, divided into two categories: general creativity methods and creativity methods for application in engineering.

### 6.2 General Creativity Methods

There are many creativity methods used in various sectors of economy. This section presents some of the widely used creativity methods taken from the published literature.[5,7]

### 6.2.1 *Crawford Slip Method (CSM)*

This creativity method was developed by C.C. Crawford in the 1920's.[7] It is quite useful to address complex problems by using a group of people. Although, CSM allows to use a group of large number of individuals, but the group usually comprised of twenty or thirty people.

In this method, the group leader devises a detailed set of target statements for eliciting responses from the group members. The members are provided with a pile of slips and are asked to write down as many ideas as possible to the set of target statements. In turn, they spend a short time in writing down their ideas using a fresh slip for each idea. At the end of this process, all slips are collected and then all the ideas are incorporated into a final report.

Some of the useful guidelines for the group participants to follow, for the purpose of making the ideas clear and concise, and easy to use at later stages, are listed below[7]:

- Avoid using jargon.
- Use simple words and short sentences.
- Write in note form.
- Do not write more than one sentence on each slip.
- Write all acronyms in full.
- Write on the top edge of each slip.
- If an explanation is considered essential for an idea, write it on a separate slip.

Additional information on the method is available in Ref. 7.

### 6.2.2 *CNB Method*

This is a group-based technique and it assumes that all the group participants clearly understand the objective of the problem and are willing to cooperate fully. The following steps are associated with this method[8,9]:

- Provide each participant of the group with a package that contains items such as problem description, a notebook, preparation material, and all relevant creative aids.
- Allow each group participant some time (e.g., one week) for finding a solution to the given problem and request that, during the allotted time period, all ideas must be recorded each and every day. At the end of the allotted time period, request all participants to choose their best ideas

and summarize the remaining ones. In addition, request them to write their thoughts in the notebook for further exploration.
- Collect all the notebooks and study them, and then prepare a report on the issue under consideration.
- Invite all group members to review the collected notebooks.
- Schedule a meeting of all involved parties for reviewing the proposed solutions and at the end choose the most promising solution.

This method is described in detail in Ref. 8.

### 6.2.3 *Force-Field Analysis Method*

This method was devised by Kurt Lewin as a model for managing change.[7] Lewin argued that a change occurs when the change driving forces exert greater pressure than the restraining forces that resist it. The tendency of people wanting the change is to push for it, in turn, this usually generates resistance. Needless to say, as per Lewin,[7] a better way to bring about change is to lower the resistance to change rather than increasing the pressure. The force-field analysis method is quite useful for identifying ways to do this. More specifically, defining the challenge, identifying strengths and weaknesses, and then doing something about it.

The following three steps are associated with the force-field analysis method[7]:

- Prepare two lists: one containing all the positive forces and the other all the negative forces.
- Develop a diagram that shows the above two lists in columns either side of a central divide. More specifically, these lists show the tug-of-war between them.
- Utilize the diagram to determine the ways and means for reducing the divide.

### 6.2.4 *Attribute Listing Method*

This is a single-person use (i.e., for problem solving/idea generation) method and it was developed by Robert P. Crawford.[7,10] The method is fairly simple and straightforward and requires the listing of attributes of an item or idea and then examining each attribute in turn with the intention of making improvements on the item/idea.

Attribute listing is particularly useful for making improvements in complex products and service procedures. For example, here one can list the stages in a service process for solving a specific problem with quality, speed, or cost. In turn, generate ideas to improve quality or speed, or reduce cost.

Finally, it is added that in this approach, the physical attributes are not the only ones that one can list but it also allows you to examine the subject under consideration from all angles. For example, one can list social attributes (e.g., responsibilities, politics, and taboos), price attributes (e.g., cost to manufacturer, customer, and supplier), process attributes (e.g., marketing, manufacturing, and distributing time), and psychological attributes (e.g., needs and image).

This method is described in detail in Refs. 7 and 10.

### 6.2.5 *The Seven by Seven Method*

This method was developed by Carl Gregory for evaluating and prioritizing a large number of ideas generated in response to a particular objective.[7] These ideas may have been generated through brainstorming or noted over a period of time as they may have come along. The name of the method is derived from the fact that one begins by setting his/her idea slips on a seven by seven racking board (i.e., the one with seven columns and seven rows).

However, one can start this method/process as soon as he/she has exhausted his/her pile of idea slips. The method is composed of the following nine steps[7]:

- Combine similar ideas.
- Discard all irrelevant ideas.
- Modify the ideas as considered appropriate.
- Defer all ideas considered untimely.
- Review all combined, excluded, modified, or deferred ideas with the aim of gaining any additional ideas or insights into the remaining slips/ideas.
- Classify ideas by dissimilar columns (i.e., put the ideas into related groups).
- Rank ideas in each and every group.
- Generalize columns by giving little to each column.
- Rank columns.

All in all, this process/method results in fewer ideas and a structured order of priority.

### 6.2.6 *Forced-Relationships Method*

This is a useful approach developed by C.S. Whiting.[10] The main objective of this method is to generate new ideas by creating a forced relationship between two or more generally unrelated ideas or items. This method is demonstrated through an example. Assume that a company manufactures items such as desk lamps, chairs, and desks. In this case, the starting point for generating a new idea could be the association between the chair and desk. From this point in time, the idea thinker would seek a line of free relationships between these two items and subsequently he/she may strike an idea for a new product.

For example, producing a new unit that combines both chair and desk. All in all, at the end, the most promising idea is selected for further investigation or implementation.

Additional information on this method may be found in Refs. 7 and 10.

### 6.2.7 *Mind Mapping Method*

This method was originally developed by Tony Buzan in the mid-1970's for note taking.[7] Also, it works extremely well for generating new ideas. Mind mapping is a visual and a free-form method of developing ideas using right-brain thinking. Mind maps make use of association literally to draw connections between ideas and develop a map of a subject. Some of the useful guidelines associated with this method are as follows:

- Make use of just keywords or if possible images.
- Begin from the center of the page and work out.
- Ensure that the center depicts a strong and clear visual image of the general theme of the map.
- Develop/create sub-centers for sub-themes.
- Put important/keywords on lines for reinforcing structure of notes.
- Print rather than writing in scripts.
- Make use of appropriate colors to depict themes and associations for making things stand out.
- Think in the term of three dimensions.
- Make use of icons, arrows, or any other visual aids for showing links between different elements.
- Avoid getting stuck in one specific area.
- Put all ideas down as they occur.

Additional information on this method can be found in Ref. 7.

### 6.2.8 *Nominal Group Technique*

This method is intended for small groups. It differs from other group methods because there is minimal interaction between group participants. The method is particularly useful when there is a definite need to minimize interaction among group members because it may result in conflict or one person is likely to dominate the group and thereby the session outcome.

The main disadvantage of this approach is that the lack of interaction removes some of the creative sparks. However, the past experiences indicate that this method works well with fairly narrowly defined problems such as developing strategy. The group decision is the final division.

The method is composed of four stages as shown in Fig. 6.1. This method is described in detail in Refs. 5 and 7.

### 6.2.9 *Rice Storm Method*

This method is also known as TKJ method and it was originally developed in Japan. The method focuses on the objective of the group rather than the group participants. Furthermore, the method preserves the anonymity of the person generating each idea. This method can be very useful in circumstances when group commitment and cohesion is required. Another main advantage of the process is that it first identify the problem and then looks for solutions. This aspect is extremely useful because misunderstanding or

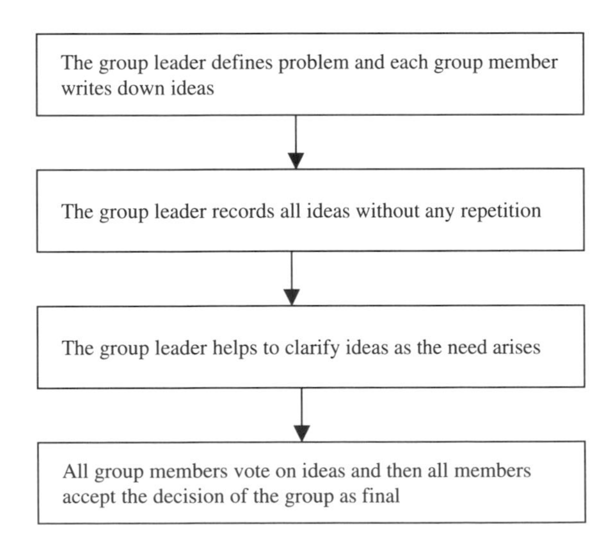


Fig. 6.1. Nominal group technique main stages.

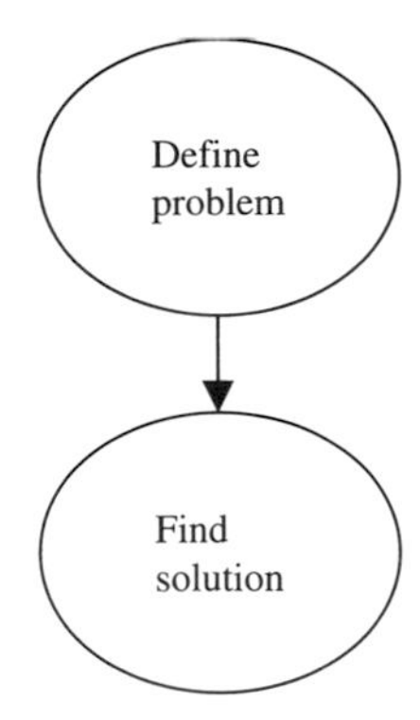


Fig. 6.2. Rice storm method stages.

disagreement about the problem can be one of the causes of failure or friction in a creative group.

Nonetheless, the rice storm method is divided into two stages as shown in Fig. 6.2.[7]

The stage "define problem" is composed of the following actions[5,7]:

- The group leader outline the general theme to the group that covers the areas of the problem. In turn, the group members write down all relevant facts on index cards (i.e., one fact per card).
- The group leader collects all the cards by ensuring their anonymity and then redistributes them to group participants.
- The group leader reads a randomly chosen card aloud.
- Group members review their stack of cards and then choose the ones that relate to the card that was read. In turn, they read out these cards to group members and at the end, they form a set of all related cards. Also, they assign a name to the set.
- This process is repeated until the grouping of all the cards into named sets.
- The same process is repeated to these sets until the creation of a single set with a name. In fact, this named set is group members' consensus definition of the problem.

Similarly, the stage "find solution" is composed of the following six actions[5,7]:

- Group participants write down each possible solution to the defined problem on a fresh index card.

- The group leader collects all the cards and redistributes them to all group members.
- The group leader selects a card at random and reads it out to the members of the group.
- The group members review all their cards and select the ones that relate to the card that was read. At the end, they form a set of all cards or ideas and then assign a name to it.
- The process is repeated until the sorting of all cards into sets with assigned names.
- A resulting solution set is generated and named from these sets. The group leader with the help of group participants turn the resulting solution set into a statement that expresses the consensus solution.

Additional information on this method is available in Refs. 5 and 7.

### 6.2.10 *Delphi Method*

Its name is derived from a city of ancient Greece called "Delphi", home to the Delphi Oracle. Here, the Oracle was approached to get information about future. Nonetheless, the Delphi method was originally developed at the RAND Corporation by Olaf Helmer and Norman Dalkey in 1969 for technical forecasting.

This method may simply be described as a group decision process about the likelihood that certain events will occur. More specifically, this is a noninteractive method that generates creative input from a disparate group of individuals. Furthermore, the Delphi method is quite specialized and is used when one wants to call on the ideas of a group of experts (who are probably geographically separated) for generating a forecast.[7]

Questionnaires are sent to each group member (i.e., expert), their inputs/responses are combined and refined, and then sent back to them. This process is repeated until the consensus is reached.

### 6.2.11 *Six Thinking Hats Method*

This is a quite useful method and it was developed by Edward deBono in the early 1980's.[7] The method sets out a framework for thinking that considers six different modes of thinking and enhances thought clarity by utilizing only one at a time. More specifically, this is a method of thinking about any problem which separates an individual's or a group's thoughts into six different types of thinking (e.g., logical, emotional, creative, etc.).

DeBono used six different colors to identify six different types of thinking (i.e., white hat thinking, red hat thinking, yellow hat thinking, black hat thinking, green hat thinking, and blue hat thinking). Each of these thinking hats is described below[7]:

- **White hat thinking.** This is concerned with pure facts and figures (i.e., looking at data without passing any judgments).
- **Red hat thinking.** This is associated with expressing feelings, intuitive responses, and hunches.
- **Yellow hat thinking.** This is a positive and optimistic thinking hat. More specifically, this is the hat of constructive thinking that examines advantages and reasons why the project/plan will work.
- **Black hat thinking.** This is the most negative hat and it examines obstacles and reasons why the item/plan/project would not work.
- **Green hat thinking.** This hat is used for the most creative thinking and it may be said that it is the hat of change, provocative ideas, and alternatives.
- **Blue hat thinking.** This hat is used when one is standing back and taking an overview. More specifically, in this case, one looks not so much at the subject but at the thinking itself.

This method is described in detail in Refs. 5 and 7.

### 6.2.12 *Assumption Reversal Method*

This is a method for the individual use and is concerned with looking at a situation from the totally opposite perspective (i.e., identify the assumptions one is making and then turn them around). The main idea behind this approach is that the original assumptions are not necessarily incorrect, but by reversing them, one generates new solutions. Furthermore, sometimes the original assumptions are indeed incorrect and false assumptions limit the range of possible solutions. Needless to say, by breaking the assumptions, one widens the scope for resolving the problem under consideration.

Often, a question is raised on what assumptions should be first reversed. Past experiences, indicate that it is generally most advantageous to reverse the most basic ones first.

An example of an assumption is: customers want good service and its reversal is: customers do not want good service. Additional information on this method is available in Ref. 7.

### 6.2.13 *Storyboarding Method*

This method was originally developed by Walt Disney for planning animated films and it was further refined by Mike Vance, one of Walt Disney's executives. Walt Disney developed series of illustrations depicting important scenes in the film, and then developed a story around each for the purpose of fleshing out the plan. In its business use, the storyboarding method creates a board for setting out important concepts and then linking them together.

Storyboarding is a very creative process for project management and problem solving and some of its benefits are as follows[7]:

- A useful tool for immersing into a project/problem, piggy-backing on ideas, and seeing new areas for attention.
- Permits the visualization of the entire picture because of putting the ideas up on a board. More specifically, this allows to see how the ideas interconnect and fit together.
- The storyboard can be kept in place on the wall throughout the life of the project or the problem-solving process.

This method is described in detail in Refs. 5 and 7.

### 6.2.14 *Two Words Method*

This is a simple and straightforward method for generating new ideas or finding solutions to a problem. The method requires to reduce a problem/subject to two words and then list synonyms for each of these two words. In the next step, the synonyms are put together in new combinations, where no two words mean exactly the same thing (i.e., they all have different nuances at the least). This way, by substituting new words in place of the original ones, one generates a new range of creative stimulants.

This method is described in detail in Ref. 7.

### 6.2.15 *Brainsketching Method*

This is a group method that harnesses the creative power of visual thinking. It is quite similar to the brainwriting approach but uses visual images instead of words. The following steps are associated with this method[7]:

- **Establish problem and draw ideas.** This is concerned with establishing the focal question and then each group participant draws an idea for a solution on a sheet of paper. The group members are prohibited to converse during this period.

- **Exchange drawings.** After a specified time period, all participants exchange their drawings with each other.
- **Study and improve drawings.** After exchanging drawings, each group participant studies the new drawing and then tries to improve it.
- **Repeat process.** The above process is repeated a number of times during the allocated time.
- **Evaluate ideas.** In this step, all group members participate to evaluate ideas on each sheet of paper.

This method is described in detail in Ref. 7.

### 6.2.16 *Excursion Method*

This is a group technique and is used in situations where other methods have failed to generate a solution. Past experiences indicate that the excursion method works quite well when the problem is narrow in scope and one requires a radical solution but has failed. The excursion approach pulls out all the stops in finding solutions to difficult problems through combining visualization and analogy methods in a group setting.

The method is composed of the following four steps[7]:

- **Define problem and go on excursion.** In this step, the group leader outlines the problem to be solved and then determines a place of a real or imaginary excursion (e.g., a mountain road, a railway journey, or a journey through space). After this, each group member closes his/her eyes and spends about ten minutes going on a private visual journey in a location outlined by the group leader. The group participants examine everything on their excursion in detail and at the end of the excursion, they document on a sheet of paper what they have seen during the excursion.
- **Establish analogies.** In this step, each group member finds analogies between the images from the excursion, noted on the sheet, and the problem at issue.
- **Evaluate the analogies.** This probably is the most crucial step where each group participant identifies the practical use of the analogies to find solutions to the problem.
- **Share with the group.** This is the final step and it introduces valuable group interactions. Each group participant shares his/her ideas

with others, starting with the excursion and the analogies, and then the proposed solutions.

Subsequently, group, as a whole, builds on these ideas to find a workable solution to the problem.

### 6.2.17 *Brainwriting Method*

This method was developed at the Batelle Institute in Frankfurt, Germany and it is basically a silent version of the brainstorming method.[7] Brainwriting approach is very useful in situations where a group is composed of some individuals who are much quieter than others and there is concern that their ideas may go unvoiced or unheard. This method allows everyone to come up with ideas simultaneously. However, it lacks a little in spontaneity.

The method is composed of the following steps[5,7]:

- **Identify and outline problem.** This is concerned with the identification of the problem to be solved and then informing it to all group members to which they must address their ideas.
- **Write down ideas.** In this step, all group members sit around a table and individually write down their ideas. This process usually takes about five minutes.
- **Pass on ideas I.** In this step, each group member passes his/her ideas sheet to the next person and then each person builds on the passed on ideas. Ultimately, he/she adds more of his/her own ideas to the sheet. This entire process normally takes about five minutes.
- **Pass on ideas II.** In this step, the idea sheets are passed on again. This passing on process continues to the point till the ideas appear to dry up.
- **Collect idea sheets.** In this step, the group leader collects all the sheets and reads out all the ideas to group members. In turn, the group evaluates ideas and selects the most appropriate ones.

The following rules are associated with this method[7]:

- Do not allow any criticism of anyone's ideas during the session.
- Encourage freewheeling during the session (i.e., the wilder the idea, the better it is).
- Defer any judgment on any idea until the evaluation process at the end of the session.

### 6.2.18 *Random Stimulation Method*

By introducing a totally random element into thinking about a challenge or problem helps to stimulate one's mind to discover new connections and patterns. Random stimulation is a simple, popular, and efficient method. It works quite well when one feels that the problem or challenge under consideration is too rigid or difficult and a fresh approach is needed.

This method, for individual use, basically requires to choose a suitable random word, usually a noun, and then think about it. There, relate your thoughts to the problem under consideration. Document your thoughts before trying another word if the need arises.

This method is described in detail in Refs. 5 and 7.

### 6.2.19 *Scenarios Method*

Scenarios are qualitatively different descriptions of plausible futures. They are used quite frequently to prepare alternative strategies for businesses. The main idea of scenarios is to develop a number of different and plausible models of key drivers for future development, such as changes in new technology, competition, and economic growth. This approach is quite useful to identify strategies for building strengths and reducing weaknesses, in order to reduce threats to the business and enhance opportunities.

Needless to say, any changing problem situation is a candidate for the application of the scenarios method. The method is composed of the following steps[7]:

- **Identify problem.** This is concerned with stating the problem to be studied. For example, "What will be the impact of developing technology on the business?"
- **Identify key drivers.** This is concerned with identifying a number of key drivers such as the economy, cash flow, competition, and new technology that will determine the organization's future.
- **Develop a future scenario.** This is concerned with constructing a future scenario around each of the above key drivers.
- **Focus on scenario outline effects.** This is concerned with focusing on how the scenario outline will affect the key drivers.
- **Summarize scenario impacts and develop strategies.** This is concerned with summarizing all scenarios and their impacts on the business, and then developing appropriate strategies.

### 6.2.20 *Lotus Blossom Method*

This is a widely used method of generating ideas in a group and it was developed by a Japanese named, Yasuo Matsumura.[5,7] The Lotus blossom method is quite useful to generate lots of ideas to work on at the beginning of a process.

The method is named after the Lotus flower because it replicates its structure. More specifically, lotus blossom petals radiate from the center. Thus, in this method, ideas radiate out from the center by very much following the same pattern. In turn, all the radiated ideas become the center of a new lotus blossom.

This method is described in detail in Ref. 7.

## 6.3 Creativity Methods for Application in Engineering

Over the years, a very large number of creativity methods have been developed.[5,7,11,12] This section presents a number of such methods which are considered useful for application in engineering areas.[1,9,12]

### 6.3.1 *Group Brainstorming Method*

This is one of the most widely used group creativity methods in the industrial sector. This method was practiced by Hindu religious teachers of India for over four centuries, and they called it Prai-Barshana (Prai means "outside yourself" and Barshana means "question"). Consequently, the modern term "brainstorming" basically means "using the brain to storm a problem". In modern context, this method was first applied by Alex Osborn in 1941 to improve advertising ideas in both quantity and quality.[7,10]

A group of people take part in brainstorming sessions. Past experiences indicate that the best results are generated when the group is composed of 8–12 individuals. The individuals participating in the sessions belong to different backgrounds but have similar interest.

During the sessions, one idea for solving the given problem triggers another idea and the process continues. Furthermore, during these sessions, the concentration on the problem under consideration is very intense with the aim of generating at least 50 ideas in each session. Usually, a brainstorming session is held for less than an hour.

Some of the useful guidelines for holding effective brainstorming sessions are listed below[7,9,12,13]:

- **Use appropriate humor.** The purpose of this action is to relax participants into a more creative frame of mind.
- **Do not allow criticism.** This means no criticism whatsoever is allowed during any brainstorming session.
- **Keep the ranks of group members fairly equal.** Past experiences indicate that it takes a considerable amount of warm-up time for low-ranking persons to mix their views freely with those put forward by more senior people.
- **Encourage free wheeling.** This basically means that the wilder the idea, the better it is, because it may spin-off other good ideas.
- **Choose the timing of the session carefully.** This basically means that ill-timed brainstorming sessions may result in lower productivity of the participants subsequent to the sessions.
- **Record ideas.** This means that record all ideas during the sessions as they are generated.
- **Think of some possible solutions to the brainstorming session problem ahead of time.** The purpose of having these solutions is to kick them in, to stimulate the idea-generation process when the participants' idea output begins to lag.
- **Combine and improve ideas.** A single idea may not be an effective solution to the problem under consideration, but combining it with other ideas may result in a better solution to the problem being pursued.

### 6.3.2 *Morphological Analysis Method*

This method is well-known in engineering design and it was developed by Fritz Zwicky in the 1960's.[1,7] It is particularly a useful approach for modifying products or services, or developing new ones.

The method may simply be described as the study of structure and form that helps to generate ideas by creating new combinations of attributes. More specifically, in this method, a given problem is divided into a number of functions (sometimes even into sub-functions) that must be carried out and alternative ideas are generated for each of these functions or sub-functions. It means that there are a large number of possible solutions to the problem, generated from the number of permutations of the possible solutions to each of the functions. This can create some difficulties to the solution seeker in choosing the best solution from the large number of available options.[14] Nonetheless, some examples of the use of this method are presented in Ref. 15.

This method is described in detail in Refs. 1 and 7.

### 6.3.3 *Synectics Method*

This method was developed by William Gordon on the premise that creative problem-solving is best accomplished through nonrational thought to reach a rational solution.[7,16] The term "synectics" is derived from the Greek language and it means bringing forth together.

In simple words, the idea behind this group problem-solving method is that one can find links between seemingly unconnected things, and by putting them together can result in a solution. Also, it may be added that synectics is a bit like the group brainstorming method with many other bits thrown in and it makes use of a range of other creativity approaches, particularly analogies and metaphors.

Three basic assumptions associated with the synectics method are as follows[7]:

- Creativity increases with the understanding of mental processes that determine our behavior.
- The irrational and emotional elements of an individual's creative behavior are more important than his/her rational and intellectual elements.
- People can learn to harness the above irrational and emotional components/elements.

The synectics process usually requires a group leader who encourages the criticism of ideas at certain times to harness the emotional response. Also, this person brings in creative techniques such as analogies and metaphors, association, and excursion into the process at the appropriate time.

The main steps of the synectics method are as follows[7]:

- Identify the problem.
- Make the familiar strange.
- Relate the ideas generated back to the original problem and then find a suitable solution to the problem.

### 6.3.4 *Checklist Method*

This is one of the simplest methods for finding solutions to a given problem. It is used often in engineering design as a means of evaluation.[1,17,18] The method is concerned with developing a list of general questions on a specified problem and then seeking appropriate answers to those questions. The method assumes that a feasible solution to the problem under consideration exists.

Some examples of the checklist method questions are listed below[7,17,18]:

- Is it possible to make the existing solution more compact?
- Can there be any other application of the current solution?
- What are the consequences if the proposed solution is taken to its extreme?
- Is it possible to rearrange parts?
- Can the solution be modified?
- Can the solution be combined with another solution to make it better?
- Is it possible to enlarge or reduce the solution?
- Can the current benefits of the existing solution be further improved?
- What are the ways for making further improvements in appearance quality, and performance of that current solution?
- What are the other scientific bases that can equally be effective for the solution in question?
- Can the drawbacks of the current solution be overcome?

### 6.3.5 *And–Also Method*

This method is use by two persons.[8,9,19] Both the persons select a topic for brainstorming, and then agree with each other's idea and add something to other's idea. This process continues until an acceptable idea is found.

### 6.3.6 *Tear-Down Method*

This method is essentially the same as the And–Also method but with one exception (i.e., instead of agreeing with each other, they totally disagree or tear-down each other's idea and then suggest a new idea).[8,9,19]

More specifically, after selecting a topic for brainstorming, person $X$ takes the attitude that the present way is incorrect and then contributes an idea suggesting another way, but not necessarily better. Person $Y$ is forbidden to agree with person $X$'s idea. Consequently, person $Y$ has to suggest another way, again not necessarily better. In turn, person $X$ is not permitted to agree with $Y$'s way/idea and has to come up with still another idea/way. This process continues until an acceptable idea is found.

### 6.3.7 *Distribution Method*

This method was developed for use by a single person.[9,19] In this case, a person documents the problem and then distributes it to a number of people prior to the meeting. Furthermore, this person clearly specifies a number of solution ideas per participant as a mandatory requirement to participate in the meeting.

This exercise permits the person to gather a large number of tentative solution ideas from the meeting participants. After a careful evaluation, one of these ideas may end up as an acceptable solution to the problem.

## Problems

1. Write an essay on creativity methods.
2. List ten most important creativity methods.
3. Discuss the group brainstorming method.
4. Compare the group brainstorming method with the brainwriting method.
5. Discuss the two most important creativity methods for use in engineering.
6. Discuss the following creativity methods:
   - Mind mapping method
   - Rice storm method
   - Assumption reversal method
7. Compare the and–also method with the tear-down method.
8. Discuss the main stages of the nominal technique.
9. Discuss the morphological analysis method.
10. What is the synectics method?

## References

1. Thompson, G. and Lordan, M., A Review of Creativity Principles Applied to Engineering Design, *Proc. Inst. Mech. Engrs.* (*British*), Vol. 213, Part E, 1999, pp. 17–31.

2. Van Gundy, A.B., *Managing Group Creativity: A Modular Approach to Problem Solving*, American Management Association, New York, 1984.
3. Hicks, M., *Problem Solving in Business and Management*, Chapman and Hall, London, 1991.
4. Geschka, H., From Experience, Creativity Workshops in Product Innovation, *Journal of Product Innovation Management*, Vol. 4, 1986, p. 1.
5. Higgins, J.M., *101 Creative Problem Solving, Techniques: The Handbook of New Ideas for Business*, New Management Publishing Company, Winter Park, Florida, 1994.
6. Schlicksupp, H., Idea Generation for Industrial Firms: Report on an International Investigation, *R & D Management*, Vol. 17, No. 2, 1977, pp. 10–14.
7. Jay, R., *The Ultimate Book of Business Creativity: 50 Great Thinking Tools, for Transforming Your Business*, Capstone Publishing Limited, Oxford, UK, 2000.
8. Haefele, J.W., *Creativity and Innovation*, Reinhold Publishing Corporation, New York, 1962.
9. Dhillon, B.S., *Engineering and Technology Management: Tools and Applications*, Artech House, Inc., Boston, 2002.
10. Osborn, A.F., *Applied Imagination*, Charles Scribner's Sons, New York, 1963.
11. Rothenberg, A. and Greenberg, B., *The Index of Scientific Writings on Creativity*, Archon Books, Hamden, Connecticut, 1976.
12. Dhillon, B.S., *Engineering Management: Concepts, Procedures, and Models*, Technomic Publishing Company, Lancaster, Pennsylvania, 1987.
13. Beakley, G.C. and Chilton, E.G., *Introduction to Engineering Design and Graphics*, McMillan, New York, 1973.
14. Whiting, C.S., *Creative Thinking*, Van Nostrand Reinhold Company, New York, 1958.
15. Cross, N., *Engineering Design Methods*, John Wiley and Sons, New York, 1989.
16. Gordon, W.J., *Synectics*, Harper and Brothers, New York, 1961.
17. Walton, J., *Engineering Design: From the Art to Practice*, West Publishing Company, New York, 1991.
18. Stoecker, W.F., *Design of Technical Systems*, McGraw-Hill Book Company, New York, 1980.
19. Studt, A.C., How to Set Up Brainstorming Sessions, in *Management Guide for Engineers and Technical Administrators*, edited by Chironis, N.P., McGraw-Hill Book Company, 1969, pp. 276–277.

## Chapter 7

# Creativity Measurement and Analysis

## 7.1 Introduction

Today, the fields of quality, reliability, safety, and human factors are well-developed disciplines of engineering. Over the years, a large amount of published literature on these areas has appeared in the form of journal articles, technical reports, conference proceedings, and books.[1,2] Many new concepts and methods have been developed in these areas for performing various types of measurement and analysis. Some examples of these concepts and methods are fault tree analysis (FTA), control charts, cause and effect diagram, and probability tree analysis.

These concepts and methods can also be applied to perform various types of creativity measurement and analysis. Nonetheless, over the years, some attempts have been made to develop a formula for estimating creative ideas and metrics for determining innovative company performance.

This chapter presents some metrics for determining innovative company performance, a formula for estimating creative ideas, and the application of some measurement and analysis concepts and methods, developed in quality, reliability, safety, and human factors areas to the creativity field.

## 7.2 Metrics for Determining Innovative Companies' Performance

Increasingly managers are faced with judging the effectiveness of organization performance with respect to creativity and innovation. It is a rather difficult task because of innovative performance's mix and intricate relationships with team-work, time lines, value perception, etc.[3] However, over the years, many metrics have been developed for use as indicators of innovative performance for a profit center or an organization/company as whole. Some of these metrics are presented in Table 7.1.[3,4] Nonetheless, the most

Table 7.1. Metrics for determining innovative companies' performance (these metrics are usually normalized in various dimensions such as number of suggestions per year and per $M sales per year).

| No. | Variable | Measure/indicator |
|---|---|---|
| 1 | Patents and publications | • Number of patents and publications<br>• Number of citations<br>• Royalties from patents |
| 2 | Response time | • Time to fill orders<br>• Time-to-market<br>• Research and development response time to in-house request |
| 3 | Cost reduction | • Cost reduction of internal business process<br>• Cost reduction of existing product or service |
| 4 | New product/service | • Number of new product/service concepts identified<br>• Price-performance of new product/service<br>• Revenue ratios: new versus old products<br>• Number of new service/product concepts introduced to market |
| 5 | Customer satisfaction | • Repeat business<br>• Referrals<br>• Critics' report<br>• Satisfaction measured through survey/other feedback |
| 6 | Business success | • Sales volume<br>• Pay-back period<br>• Media coverage<br>• Market share<br>• Profitability measures<br>• Corporate score card<br>• World leadership image |

commonly used of these metrics are as follows[3,4]:

- Time-to-market (products/services).
- Number for new products/services introduced to market.
- Patent disclosures.
- Cost and performance improvements.

It is to be noted that the above four items/measures/indicators are usually normalized in various dimensions (e.g., number of suggestions per year, per \$M sales per year, and per employee per year).

The meaningful performance measures with respect to creativity and innovation are even more difficult to define for task teams and individual contributors. However, some of the common metrics used for this very purpose are as follows[3,4]:

- Number of innovative ideas.
- Number of patent disclosures and papers.

## 7.3 A Formula for Predicting Creative Ideas

Sometime, the need may arise to predict the total number of ideas to be generated by a group of population. Reference 5 has proposed the following formula for this very purpose:

$$\theta(T) = \int_0^T \int_0^\alpha p(t) nt \, dn dt \,, \tag{7.1}$$

where

$\theta(T)$ is the total number of creative ideas.
$n$ is the number of creative ideas per person in time unit $t$.
$\alpha$ is the maximum number of ideas.
$p(t)$ is the population at time $t$.

## 7.4 Fault Tree Analysis (FTA)

This method was developed in the early 1960's at the Bell Telephone Laboratories to perform reliability and safety analyses of Minuteman Launch Control System. It can also be used to perform various types of creativity-related analysis.

A fault tree is a logical representation of basic events that may cause the occurrence of a specified undesirable event, often known as the "Top Event".

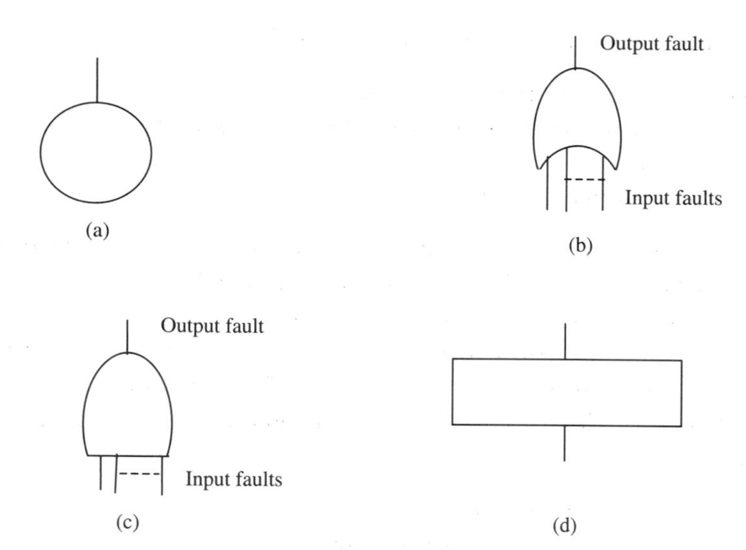


Fig. 7.1. Common fault tree symbols: (a) Circle, (b) OR gate, (c) AND gate, and (d) Rectangle.

In the construction of a fault tree, various symbols are used. Figure 7.1 presents four commonly used symbols in the construction of fault trees. These four symbols are described below[6,7]:

- **Circle.** This denotes a basic fault event (e.g., the failure of an elementary component). The specific values of this event's parameters such as occurrence rate and occurrence probability are normally obtained from empirical data.
- **Rectangle.** This represents a fault event that results from the logical combination of fault events through the input of a logic gate such as OR and AND.
- **OR Gate.** This denotes that an output fault occurs only if any one or more input fault events occur.
- **AND Gate.** This denotes that an output fault event occurs only if all the input fault events occur.

**Example 7.1** Assume that middle level managers of a well-established company have a very creative idea about developing a new product. This new product idea can fail due to three factors: lack of higher management interest, lack of sufficient funds, and insufficient time allocation. In turn, two causes for the lack of interest of the higher management are the lack of proper presentation and too many other new ideas in pipeline. Similarly, two causes for the lack of sufficient funds are the under-estimation of the required amount and funds diverted to other areas.

Develop a fault tree for the occurrence of the top event "new product idea failure" by using Fig. 7.1 symbols.

A fault tree for the example is shown in Fig. 7.2. The single capital alphabet letters in the figure denote the corresponding fault events (e.g., $T$: new product idea failure, $Y$: insufficient time allocation, and $Z$: lack of sufficient funds).

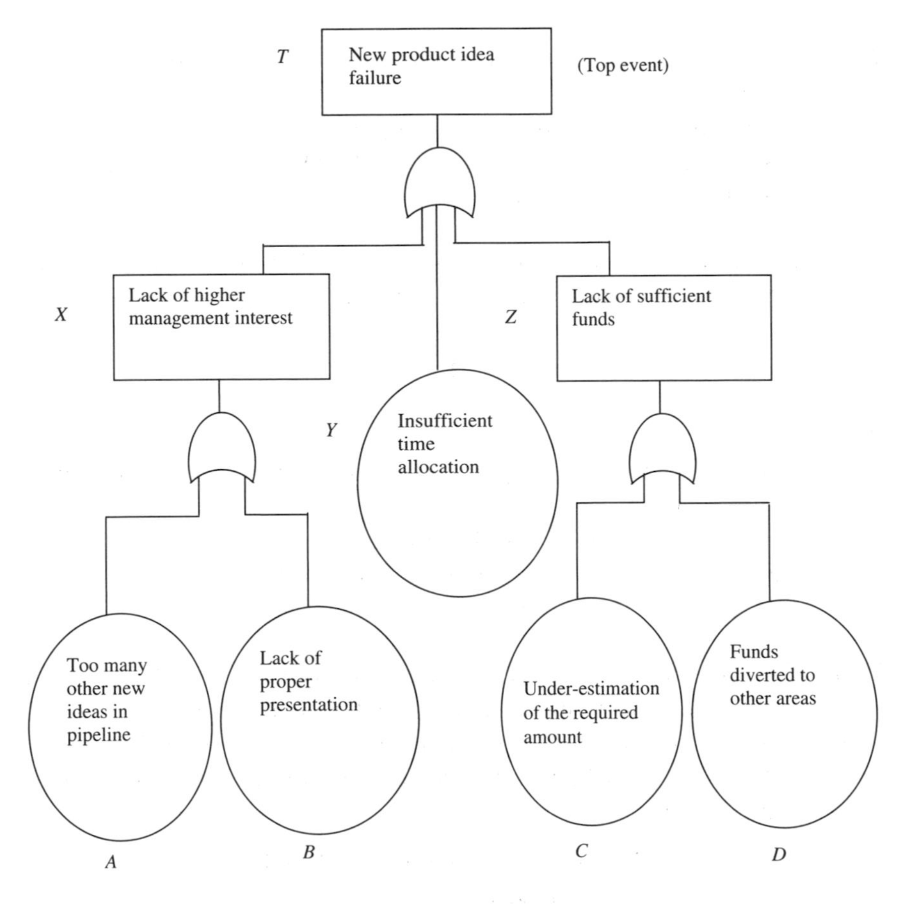


Fig. 7.2. A fault tree for Example 7.1.

### 7.4.1 *Fault Tree Probability Evaluation*

The probability of occurrence of the top event (e.g., $T$ in Fig. 7.2) can be calculated if the occurrence probabilities of basic faults (e.g., $A$, $B$, $C$, and $D$ in Fig. 7.2) are known. This can only be calculated by first estimating the occurrence probabilities of the output fault events of lower and intermediate logic gates such as OR and AND. The occurrence probability of the OR gate output fault event can be estimated by using the following expression[8]:

$$P(X_0) = 1 - \prod_{i=1}^{n} \{1 - P(X_i)\}\,, \tag{7.2}$$

where

$P(X_0)$ is the OR gate output fault event, $X_0$, the probability of occurrence.

$n$ is the total number of OR gate input fault events.

$P(X_i)$ is the probability of occurrence of the OR gate input fault event $X_i$; for $i = 1, 2, 3, \ldots, n$.

Similarly, the occurrence probability of the AND gate output fault event can be estimated by using the following expression[8]:

$$P(Z_0) = \prod_{i=1}^{m} P(Z_i)\,, \tag{7.3}$$

where

$P(Z_0)$ is the AND gate output fault event, $Z_0$, the probability of occurrence;

$m$ is the total number of AND gate input fault events; and

$P(Z_i)$ is the probability of occurrence of the AND gate input fault event $Z_i$; for $i = 1, 2, 3, \ldots, m$.

**Example 7.2** Assume that in Fig. 7.2, the occurrence probabilities of events $A$, $B$, $C$, $D$, and $Y$ are 0.002, 0.003, 0.004, 0.005, and 0.001, respectively. Calculate the probability of occurrence of the top event $T$.

Using the given probability values for the occurrence of events $A$ and $B$ in Eq. (7.2), we get

$$\begin{aligned} P(X) &= 1 - (1 - 0.002)(1 - 0.003) \\ &= 0.0049\,, \end{aligned}$$

where $P(X)$ is the probability of occurrence of fault event $X$.

Similarly, using the given data values for the occurrence of fault events $C$ and $D$ in Eq. (7.2) yields

$$P(Z) = 1 - (1 - 0.004)(1 - 0.005) \\ = 0.0089\,,$$

where $P(Z)$ is the probability of occurrence of fault event $Z$.

By inserting the above two calculated values and the given value for the occurrence of fault event $Y$ into Eq. (7.2), we get

$$P(T) = 1 - (1 - 0.0049)(1 - 0.0089)(1 - 0.001) \\ = 0.0147\,.$$

Thus, the probability of occurrence of the top event $T$ is 0.0147.

## 7.5 Control Charts

These charts were developed by Walter A. Shewhart in 1924 for application in quality control work.[9] A control chart may simply be described as a graphical approach used for determining if a process is in a "state of statistical control" or out of control.[10] In the area of creativity and innovation, the process could be the frequency of new ideas, successful implementation of new ideas, and so on.

Although there are many different types of control charts, their application to creativity-related problems is demonstrated through only one type of control charts known as the $C$-Charts.[11,12] The expressions for the control limits of this type of control charts are based on the Poisson distribution. Thus, the mean of the Poisson distribution with respect to creative ideas is expressed by[13]:

$$\theta = \frac{N_c}{T}\,, \tag{7.4}$$

where

$\theta$ is the mean value of the Poisson distribution.
$T$ is the total number of items or time period.
$N_c$ is the number of creative ideas.

The standard deviation of the Poisson distribution is given by[13]:

$$\sigma = (\theta)^{1/2} . \tag{7.5}$$

Thus, the upper and lower control limits of the $C$-Chart are[13]:

$$\text{UCL} = \theta + 3\sigma \tag{7.6}$$

and

$$\text{LCL} = \theta - 3\sigma , \tag{7.7}$$

where

LCL is the lower control limit of the $C$-Chart.
UCL is the upper control limit of the $C$-Chart.

These two equations are based on the assumption that there is a 99.73% chance that a random sample will fall within the values generated by them.

**Example 7.3** Assume that employees of an engineering company during a 12-month period generated a total of 85 new ideas and their monthly breakdowns are presented in Table 7.2. Develop the $C$-Chart and comment on the end result.

Table 7.2. New ideas generated over a 12-month period in an engineering company.

| No. | Month | No. of new ideas |
|---|---|---|
| 1 | January (J) | 6 |
| 2 | February (F) | 9 |
| 3 | March (MR) | 10 |
| 4 | April (A) | 12 |
| 5 | May (M) | 8 |
| 6 | June (JN) | 5 |
| 7 | July (JU) | 6 |
| 8 | August (AT) | 4 |
| 9 | September (S) | 7 |
| 10 | October (O) | 3 |
| 11 | November (N) | 11 |
| 12 | December (D) | 4 |

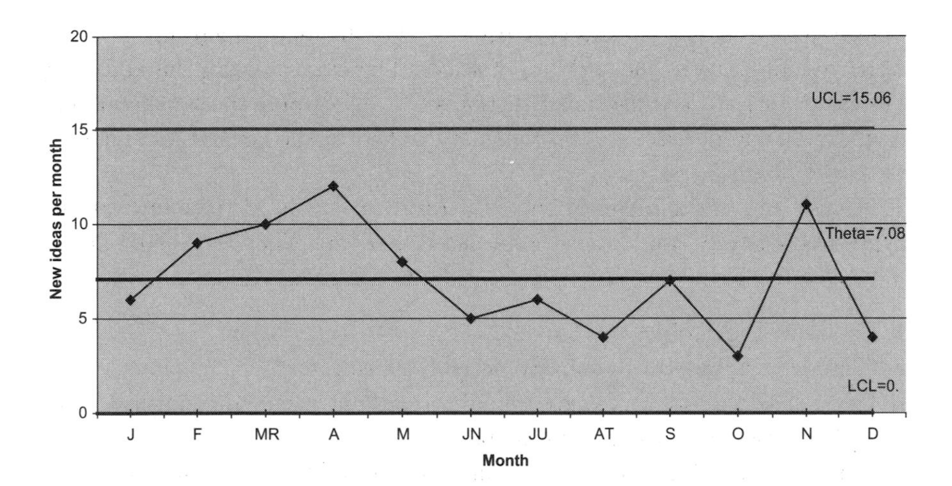


Fig. 7.3. The $C$-chart for Example 7.3.

Using the given data values in Eq. (7.4) yields

$$\theta = \frac{85}{12} = 7.08\,\text{ideas/month}\,.$$

Thus, from Eq. (7.5), we get

$$\begin{aligned}\sigma &= (7.08)^{1/2} \\ &= 2.66\,.\end{aligned}$$

Using the above calculated values in Eqs. (7.6) and (7.7), we get

$$\text{UCL} = 7.08 + 3(2.66) = 15.06$$

and

$$\text{LCL} = 7.08 - 3(2.66) = -0.9\,.$$

The $C$-Chart for the above-calculated values and Table 7.2 data values is shown in Fig. 7.3. It is to be noted that as the value of LCL $= -0.9$, in the figure, it was taken as 0.0. Nonetheless, the chart shows that the monthly generation of new ideas is within the control limits; thus the idea generation process is in control.

## 7.6 Cause and Effect Diagram

This is a widely used method in quality control work. It was developed in the early 1950's by K. Ishikawa, a Japanese quality expert.[14] This method is also known as the Ishikawa diagram or the "fishbone diagram" because of its resemblance to the skeleton of a fish. The extreme right-hand side of the diagram (i.e., the fish head) denotes the effect and the left-hand side displays all the possible causes which are linked to the central line referred to as the "Fish Spine".

Cause and effect diagram could be quite useful in performing various types of creativity- and innovation-related analysis. The steps listed below are followed in constructing a cause and effect diagram[14]:

- Establish the problem statement.
- Conduct brainstorm to identify possible causes.
- Develop the major cause groups by stratifying them into natural categories/classes and the process steps.
- Construct the diagram by connecting the highlighted causes under appropriate process steps and fill the effect (i.e., the problem under consideration) in the diagram box (i.e., the fish head) located at the extreme right-hand side of the diagram.
- Refine the cause categories by asking questions such as follows:
  - What is the main reason for the existence of this condition?
  - What causes this?

Some of the important benefits of the cause and effect diagram are useful to generate relevant ideas, useful to identify root causes, useful to present an orderly arrangement of theories, and an useful guide to further inquiry.

**Example 7.4** Inspection personnel of an electrical cable manufacturing company have identified a pressing problem (i.e., cable insulation is too thick) with cables being manufactured. Develop a cause and effect diagram for this problem.

After a careful investigation, three major cause areas for the problem were identified: machine ($M$), insulation material ($IM$), and die ($D$). Sub-causes for each of these cause areas were brainstormed and identified. They are presented in Table 7.3.[15] The cause and effect diagram for the example is shown in Fig. 7.4.[15]

## 7.7 Probability Tree Analysis

This is a quite useful method for conducting task analysis by diagrammatically representing critical human actions and other associated events.[16] Diagrammatic task analysis is denoted by the branches of the probability

Table 7.3. Major cause areas and their subcauses for the Example 7.4 problem.

| No. | Major cause area | Subcauses |
|---|---|---|
| 1 | Machine ($M$) | • Pressure<br>• Injection speed<br>• Ejection method |
| 2 | Insulation material ($IM$) | • Pallet size<br>• Temperature<br>• Heating rate<br>• Composition |
| 3 | Die ($D$) | • Material<br>• Cooling rate<br>• Temperature<br>• Dimension<br>• Lubricant |

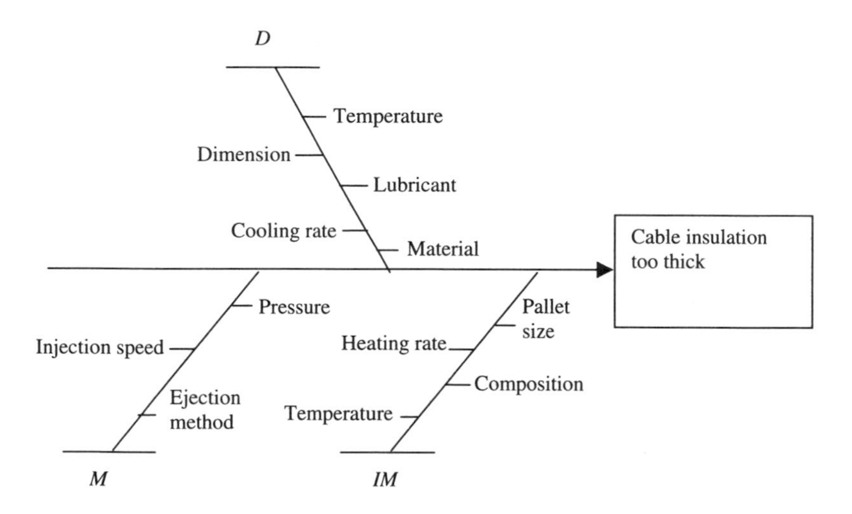


Fig. 7.4. Cause and effect diagram for Example 7.4.

tree. More specifically, the branching limbs of the tree represent the outcome of each event (e.g., success or failure) and each branch is assigned probability of occurrence.

Some of the advantages of this method are an effective visibility tool, simplified mathematical calculations, and the flexibility for incorporating (i.e., with some modifications) factors such as interaction stress, emotional stress, and interaction effects. Additional information on the method is

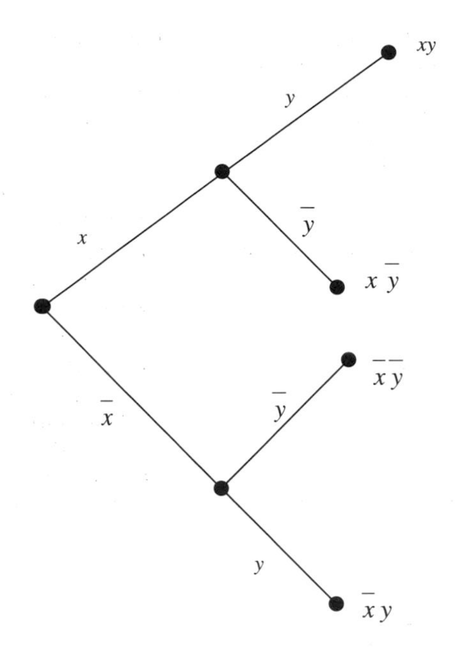


Fig. 7.5. Probability (i.e., event) tree for Example 7.5.

available in Refs. 16 and 17. The following two examples demonstrate the basics of the method.

**Example 7.5** An organization formed two independent groups $x$ and $y$ for the task of generating new creative product ideas. More specifically, the function of group $x$ is to come up with a new idea (i.e., cost effective or noncost effective) and group $y$ is to refine it. Group $y$ can perform its assigned task successfully or unsuccessfully. Develop probability (i.e., event) tree and then obtain an expression for the probability of not successfully accomplishing the overall mission (i.e., not coming up with a new cost effective and refined creative product idea).

In this example, group $x$ either comes up with a new cost effective or noncost effective idea. Similarly, group $y$ either successfully refine the new idea or fails to refine it. This scenario is depicted by the probability tree shown in Fig. 7.5.

The alphabetic symbols used in Fig. 7.5 are defined below:

$x$ denotes the event that group $x$ has generated a new cost effective idea.

$\bar{x}$ denotes the event that group $x$ has generated a new but noncost effective idea.

$y$ denotes the event that group $y$ has successfully refined the new idea.

$\bar{y}$ denotes the event that group $y$ was unsuccessful to refine the new idea.

In Fig. 7.5, events "$xy$" denote the overall mission success. Thus, the probability of occurrence of independent events "$xy$" is[16,17]:

$$P(xy) = P_x P_y\,, \tag{7.8}$$

where

$P(xy)$ is the probability of group $x$ producing a new cost effective idea and group $y$ successfully refining it.

$P_x$ is the probability of group $x$ generating a new cost effective idea.

$P_y$ is the probability of group $y$ successfully refining the new cost effective idea.

Similarly, in Fig. 7.5, events $x\bar{y}, \bar{x}\bar{y}$, and $\bar{x}y$ denote the three distinct possibilities of not having a new cost effective and refined creative idea. Thus, for independent events, the probability of not having a new cost effective and refined creative idea is:

$$P(x\bar{y} + \bar{x}\bar{y} + \bar{x}y) = P_x P_{\bar{y}} + P_{\bar{x}} P_{\bar{y}} + P_{\bar{x}} P_y\,, \tag{7.9}$$

where

$P_{\bar{x}}$ is the probability of group $x$ generating a noncost effective idea.

$P_{\bar{y}}$ is the probability of group $y$ failing to refine the new idea.

$P(x\bar{y} + \bar{x}\bar{y} + \bar{x}y)$ is the probability of not having a new cost effective and refined creative idea.

**Example 7.6** Assume that in Example 7.5 the probability of group $x$ generating a cost effective idea is 0.6. Similarly, the probability of group $y$ successfully refining it is 0.7.

Calculate the probability of successfully accomplishing the overall mission (i.e., having the cost effective idea and its successful refinement).

By substituting the specified data values into Eq. (7.8), we get

$$P(xy) = (0.6)(0.7) = 0.42\,.$$

Thus the probability of successfully accomplishing the overall mission is 0.42.

## 7.8 Creativity Improvement with Parallel Redundancy

Parallel redundancy is widely used process to improve reliability of engineering systems. It basically calls for multiple active units/items/humans to perform the same function instead of having only one unit/item/human. This concept can equally be used to improve creativity efforts in organizations.

The probability of success of independent units working in parallel is given by[17,18]:

$$P_s = 1 - (1 - P_1)(1 - P_2)(1 - P_3)(1 - P_n)\,, \tag{7.10}$$

where

$n$ is the total number of units/items.
$P_i$ is the probability of success of unit/item $i$; for $i = 1, 2, 3, \ldots, n$.

**Example 7.7** A company has two parallel independent groups ($A$ and $B$) for generating new ideas for product development. For a given time period, group $A$'s probability of success in generating a new idea is 0.75 and group $B$'s is 0.70. Calculate the probability that the company will have a new idea for product development. Comment on the end result.

Using the specified data values in Eq. (7.10) yields

$$P_s = 1 - (1 - 0.75)(1 - 0.70) = 0.925\,.$$

Thus, the probability that the company will have a new idea for product development is 0.925. If the company had only either group $A$ or group $B$, its probability of having a new idea for product development would have been 0.75 or 0.70, respectively. More specifically, by having two groups, the company improved its chances of having a new idea for product development to 0.925.

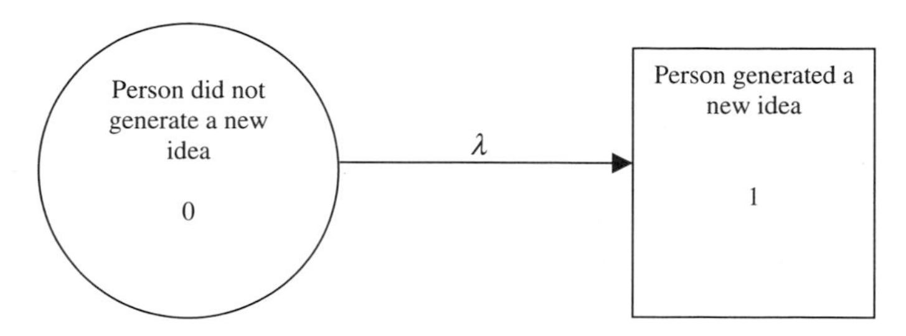


Fig. 7.6. State space diagram representing a creative person. Numerals in circle and box denote states.

## 7.9 Time-Dependent Creativity Analysis with Markov Method

Markov method is widely used in performing reliability analysis of engineering systems. The method is named after a Russian mathematician named Andrei Andreyevich Markov (1856–1922). It can be used to perform various types of time-dependent creativity analysis. The following assumptions are associated with the Markov method[19]:

- All occurrences are independent of each other.
- The probability of the occurrence of a transition from one state to another in the finite time interval $\Delta t$ is given by $\lambda \Delta t$, where $\lambda$ is the transition rate from one state to another.
- The transitional probability of more than one occurrences in time interval $\Delta t$ from one state to another is negligible (e.g., $(\lambda \Delta t)\ (\lambda \Delta t) \rightarrow 0$).

The following example demonstrates the application of this method in performing time-dependent creativity analysis.

**Example 7.8** A person generates new ideas at a constant rate $\lambda$. More specifically, a state space diagram describing this scenario is shown in Fig. 7.6. Develop probability expressions for the person generating and not generating new ideas at time $t$ by using the Markov method.

Using the Markov method, we write down the following equations for Fig. 7.6[19]:

$$P_0(t + \Delta t) = P_0(t)(1 - \lambda \Delta t)\,, \tag{7.11}$$

$$P_1(t + \Delta t) = P_1(t) + P_0(t)\lambda \Delta t\,, \tag{7.12}$$

where

| | |
|---|---|
| $t$ | is time. |
| $\lambda$ | is the constant idea generation rate of the creative person. |
| $\lambda\Delta t$ | is the probability of idea generation by the creative person in finite time interval $\Delta t$. |
| $P_0(t)$ | is the probability that the creative person did not generate a new idea (i.e., state 0 in Fig. 7.6) at time $t$. |
| $P_1(t)$ | is the probability that the creative person has generated a new idea (i.e., state 1 in Fig. 7.6) at time $t$. |
| $(1-\lambda\Delta t)$ | is the probability of no idea generation by the creative person in finite time interval $\Delta t$. |
| $P_1(t+\Delta t)$ | is the probability that the creative person has generated a new idea (i.e., state 1 in Fig. 7.6) at time $(t+\Delta t)$. |
| $P_0(t+\Delta t)$ | is the probability that the creative person did not generate a new idea (i.e., state 0 in Fig. 7.6) at time $(t+\Delta t)$. |

Using Eqs. (7.11) and (7.12), we obtain[8,20]:

$$\frac{dP_0(t)}{dt} + \lambda P_0(t) = 0\,, \tag{7.13}$$

$$\frac{dP_1(t)}{dt} - \lambda P_0(t) = 0\,. \tag{7.14}$$

At time $t = 0$, $P_0(0) = 1$ and $P_1(0) = 0$.

Solving Eqs. (7.13) and (7.14) by using Laplace transforms, we get

$$P_0(s) = \frac{1}{s+\lambda} \tag{7.15}$$

and

$$P_1(s) = \frac{1}{s(s+\lambda)}\,, \tag{7.16}$$

where $s$ is the Laplace transform variable.

Taking the inverse Laplace transforms of Eqs. (7.15) and (7.16), we get

$$P_0(t) = e^{-\lambda t}\,, \tag{7.17}$$

$$P_1(t) = 1 - e^{-\lambda t}\,. \tag{7.18}$$

Equations (7.17) and (7.18) are the probability expressions for the person not generating and generating new ideas at time $t$, respectively.

**Example 7.9** Assume that the new idea generation rate of a person is 0.1 ideas per month. Calculate the probabilities of the person generating and not generating new ideas during a 12-month period.

By substituting the given data values into Eqs. (7.17) and (7.18), we get

$$P_0(12) = e^{-(0.1)(12)}$$
$$= 0.3012$$

and

$$P_1(12) = 1 - e^{-(0.1)(12)}$$
$$= 0.6988\,.$$

It means that there is approximately 70% chance that the person will come out with a new idea during the specified time period. In contrast, there is about 30% probability that the person will fail to generate a new idea during the same time span.

## Problems

1. List at least ten metrics used for determining innovative companies' performance.
2. Write an essay on fault tree analysis.
3. Define the following two items with respect to fault tree analysis:
   - OR gate
   - AND gate
4. Assume that in Fig. 7.2, the occurrence probabilities of events $A$, $B$, $C$, $D$, and $Y$ are 0.01, 0.02, 0.03, 0.06, and 0.009, respectively. Calculate the probability of occurrence of the top event $T$.
5. Write an essay on quality control charts.
6. Discuss the following terms:
   - Upper control limit
   - Lower control limit
   - The $C$-charts
7. Assume that the employees of a company during a 10-month period generated a total of 80 new ideas and their monthly breakdowns are presented in Table 7.4. Develop the $C$-chart and comment on the end result.
8. Discuss the cause and effect diagram with respect to its application to creativity-related problems.
9. A company has three independent and parallel groups ($A$, $B$, and $C$) for generating new ideas for product development. For a given time period,

Table 7.4. New ideas generated over a 10-month period.

| No. | Month | No. of new ideas |
|---|---|---|
| 1 | March | 5 |
| 2 | April | 8 |
| 3 | May | 10 |
| 4 | June | 7 |
| 5 | July | 6 |
| 6 | August | 4 |
| 7 | September | 13 |
| 8 | October | 8 |
| 9 | November | 9 |
| 10 | December | 10 |

each group's probability of success in generating new ideas for product development is 0.60. Calculate the probability that the company will have a new idea for product development.

10. Assume that the new idea generation rate of a person is 0.02 ideas per month. Calculate the probabilities of the person generating and not generating new ideas during a 9-month period.

## References


1. Dhillon, B.S., *Reliability and Quality Control: Bibliography on General and Specialized Areas*, Beta Publishers, Inc., Gloucester, Ontario, Canada, 1992.
2. Dhillon, B.S., *Engineering Safety*, World Scientific Publishing, Inc., River Edge, New Jersey, 2003.
3. Thamhain, H.J., Managing Innovative R and D Teams, *R and D Management*, Vol. 33, No. 3, 2003, pp. 297–311.
4. Thamhain, H.J., Criteria for Effective Leadership in Technology-Oriented Project Teams, in *The Frontiers of Project Management Research*, edited by Slevin, A., Cleland, I. and Pinto, D., Project Management Institute, Newton Square, Pennsylvania, 2002, pp. 259–270.
5. Navin, F.P.D., Engineering Creativity-Doctum Ingenium, *Canadian Journal of Civil Engineering*, Vol. 21, 1994, pp. 499–511.
6. Dhillon, B.S. and Singh, C., *Engineering Reliability: New Techniques and Applications*, John Wiley and Sons, New York, 1981.
7. Fault Tree Handbook, Report No. NUREG-0492, U.S. Nuclear Regulatory Commission, Washington, D.C., 1981.
8. Dhillon, B.S., *Design Reliability: Fundamentals and Applications*, CRC Press, Boca Raton, Florida, 1999.

9. Statistical Quality Control Handbook, published by AT & T Technologies, Indianapolis, Indiana, 1956.
10. Feigenbaum, A.V., *Total Quality Control*, McGraw Hill Book Company, New York, 1983.
11. Buffa, E.S., *Operations Management: Problems and Models*, John Wiley and Sons, New York, 1972.
12. Smith, G.M., *Statistical Process Control and Quality Improvement*, Prentice Hall, Inc., Upper Saddle River, New Jersey, 2001.
13. Ireson, W.G. (ed.), *Reliability Handbook*, McGraw Hill Book Company, New York, 1966.
14. Mears, P., *Quality Improvement Tools and Techniques*, McGraw Hill, Inc., New York, 1995.
15. Tomas, S., Creative Problem-Solving: An Approach to Generating Ideas, *Proceedings of the International Conference of the Educational Society for Resource Management*, 1997, pp. 450–455.
16. Swain, A.D., A Method for Performing a Human-Factors Reliability Analysis, Report No. SCR-685, Sandia Corporation, Albuquerque, New Mexico, August 1963.
17. Dhillon, B.S., *Design Reliability: Fundamentals and Applications*, CRC Press, Boca Raton, Florida, 1999.
18. Dhillon, B.S., *Engineering Management: Concepts, Procedures, and Models*, Technomic Publishing Company, Lancaster, Pennsylvania, 1987.
19. Shooman, M.L., *Probabilistic Reliability: An Engineering Approach*, McGraw Hill Book Company, New York, 1968.
20. Dhillon, B.S., *Human Reliability and Error in Medical System*, World Scientific Publishing, River Edge, New Jersey, 2003.

# Chapter 8

# Creativity Climate

## 8.1 Introduction

Climate plays an instrumental role in the creativity of individuals, groups, or organizations. Unfortunately, it is often poorly understood or ignored altogether. It affects organizational and psychological processes, such as problem solving, conflict handling, decision making, learning and motivation, and communication.[1,2] Thus, it influences the efficiency and productivity of the organization, its ability to innovate, and the job satisfaction and the well-being of the organization members.

More specifically, the individual organization member is affected by the climate as a whole as well as by the general psychological atmosphere. There is no single separate event that generates more lasting influence on the behavior and feelings than the daily exposure to a specific psychological atmosphere. Needless to say, climate has a major impact on an individual's and organization's abilities to create and innovate. In turn, it dictates their success or failure in the competitive economic environment.

This chapter presents various important aspects of the creativity climate.

## 8.2 Variables Influencing Peoples' Perception of the Working Climate, Examples of Changes in the Total Environment Influencing Innovation, and Key Reasons for Organizations to Foster Creativity and Innovation

Over the years, professionals working in the area of creativity have observed many variables that influence peoples' perceptions of the working climate. Some of these variables are the leadership, the levels of job satisfaction, the organizational vision, goals, mission, and strategies, the personnel policies (particularly rewards and promotion), the organizational structures and systems, the personalities of the people in the organization, the amount of available monetary and physical resources, the beliefs and values of the organization, and the concerns for profits and losses.[2] In particular, it must be noted that various climate research studies indicate that the most influential variable that influences people's perceptions of the working climate is leadership. It affects somewhere between 40% and 80% of peoples' perceptions of the climate.[2]

As innovation is a fragile process, rapid or significant changes in total environment (corporate) can damage it severely, if it is not managed in an effective manner. Some examples of changes in the total environment that can directly or indirectly influence innovation are shown in Fig. 8.1.[3]

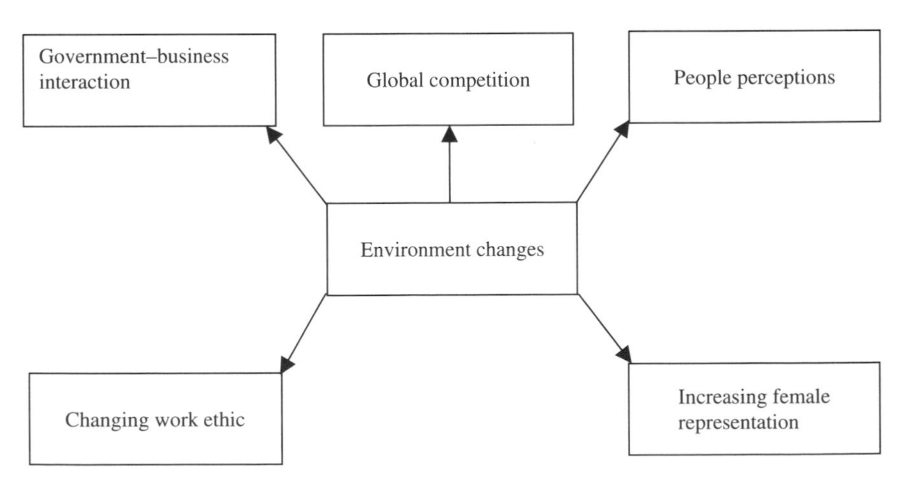


Fig. 8.1. Some examples of changes in the total environment that can influence innovation.

The government–business interaction is concerned with the increasing government regulation of business leading to disturbing changes in the general business environment. This can influence innovation to a certain degree. The global competition is concerned with corporations competing in world markets. It requires the performance of necessary risk–benefit analysis on a continuing basis to determine the proper balance between satisfying desirable social goals and achieving the appropriate economic competitiveness. Otherwise, the funding of high social goals can drain available resources away from technological innovation.

The people perceptions are concerned with factors such as inflation, energy shortages, pollution, and urban blight and strife. Usually, people look for solutions to two institutions: government and big business. Needless to say, corporate management may find itself being forced toward taking more conservative and risk-free approaches. Consequently, this may result in damage to the innovation process.

The increasing female representation is concerned with the impact of more women in labor force, particularly in management and scientific area, in the male-dominated technological industries. The effect of women on innovation directly or indirectly in the companies in such industries could be quite significant. The changing work ethic is concerned with the traditional work ethic undergoing a significant change. Particularly, items such as job satisfaction, discretionary time, and freedom of dissent are becoming important in the hearts and minds of the labor force. This can influence innovation quite significantly as management become under pressure to sacrifice profits to meet these needs of its employees.

Reference 4 has identified the following five key reasons for organizations to foster creativity and innovation:

- Superior long-term financial performance.
- Customer demand.
- Competitor copying.
- New technologies enable innovation.
- What used to work in the past, does not anymore.

Past experiences indicate that innovation is the key factor for the superior long-term financial performance of many companies. For example, as per Ref. 5, the companies that are rather successful in building wealth for their shareholders over the period of time stress innovation as an important corporate value. Nowadays, customers experience new ideas and new technologies daily and they expect innovation to a certain degree in all goods and services they use.[4,6] Competitors are getting better at copying others' innovations. New innovations are quickly reverse-engineered and replicated. Thus, in order to stay ahead in the market place, continuous innovation is definitely needed.[7]

As new technologies enable innovation, successful companies exploit innovative technology to their advantage. Needless to say, if a company does not, it may loose its market share.[8] What used to work in the past does not work now anymore because of increasing complexity and interconnectedness require a fresh approach to old problems.[9]

## 8.3 Organization's Creative Culture Attributes

Past experiences indicate that the existence of creative organizational culture is an essential element in fostering creativity. Such culture must encourage and reward employees for thinking creatively. Nonetheless, the important attributes of a creative culture are shown in Fig. 8.2.[10,11] Furthermore, the inventor of microchip, Jack Kilby, specifically recommends

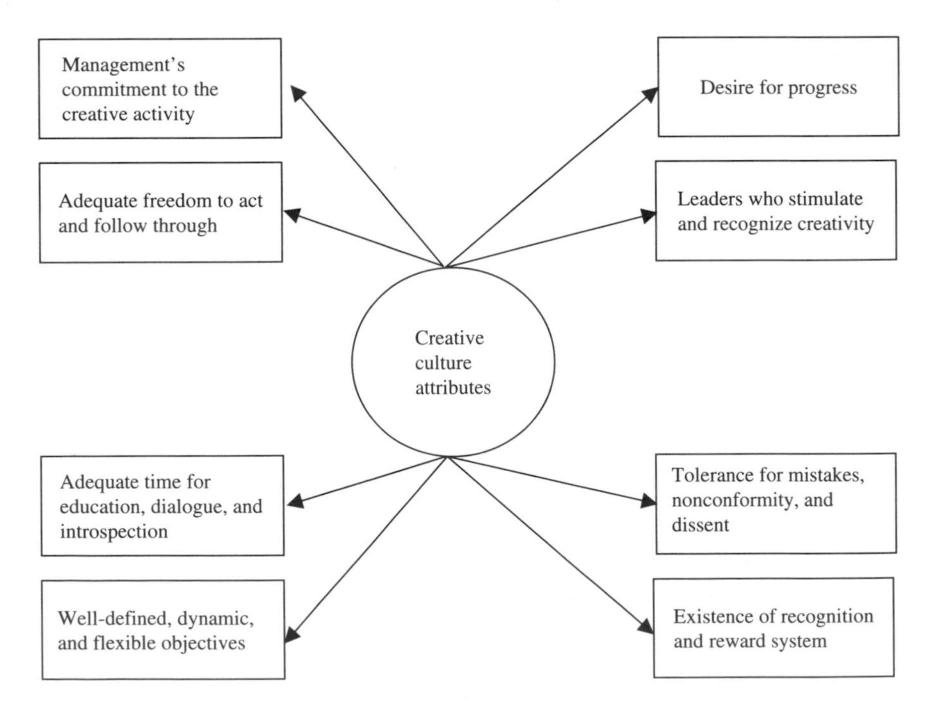


Fig. 8.2. Creative organizational culture attributes.

the following two elements for a creative culture[11,12]:

- It allows investigators to perform personal research that may benefit the company directly or indirectly.
- It requires no apology from researchers for coming up with ideas that company management did not seek at all.

## 8.4 Creative Climate Dimensions and Creative Work Environment Determinents

There are many dimensions of creative climate. They are (along with each dimension's corresponding creative climate in parentheses) challenge (energetic and enjoyable), freedom (independent initiatives), risk taking (act on new ideas), dynamism (excitedly busy), support (people listen), openness (trusting, failure accepted), debates (contentious ideas voiced), playfulness (happy, humorous), conflict (debated with insight), and idea time (off task play).[1,2,13]

According to Refs. 14 and 15, the determinants of the creative work environment can be classified under the following eight categories:

- **Organizational encouragement.** This includes determinants such as risk taking, support and evaluation of ideas, shared vision, recognition of ideas, and collaborative idea flow.
- **Organizational impediments.** This includes determinants such as rigid formal structures, internal political problems, conservatism, and destructive internal competition.
- **Freedom.** This includes determinants such as choice on how to accomplish tasks, relatively high degree of autonomy, and control over work.
- **Supervisory encouragement.** This includes determinants such as supervisory support of ideas, goal clarity, and open interactions between supervisors and subordinates.
- **Work group supports.** This includes determinants such as constructive criticism of ideas, intrinsic motivation, and backgrounds of individuals.
- **Sufficient resources.** In this case, an important determinant is adequate resource allocation. Also, perceptions of adequate resources increase creativity.
- **Challenging work.** In this case, the determinant is assignment of challenging work.
- **Workload pressures.** In this case, it may be added that although some degree of pressure has a positive influence on creativity, the extreme pressure undermines creativity.

## 8.5 Steps for Fostering Creative Environment in Companies and Guidelines for Managing Team Members that Foster Creative Work Climate

Creative environment in companies can be fostered, directly or indirectly, through the following eight steps[16]:

- Determine the specific areas where risk is considered a factor in growth. This is concerned with establishing a special set of ground rules for the identified areas of the business.
- Examine corporate policies with respect to pre-invention disclosure agreements. Limit these agreements strictly to company-related business or to the direct work assigned to an individual. Under the doubtful scenario, examine the patent laws used in countries such as United Kingdom, Germany, and Japan.
- Establish appropriate corporate long-range objectives for identified areas of internal growth. Furthermore, fund such areas on a coherent industry basis.
- Use concepts such as net present value, internal rate of return, or profitability index to calculate investment opportunities in new activities. Use these calculations primarily as a guide and remember that successful entrepreneurs always have a gut feel about new opportunities of growth.
- Develop a proper balance between regulation and development activities with respect to allocation of funds.
- Develop a proper balance among the team members with respect to factors such as academic achievements, age, and experience. For example, various studies indicate that creativity declines with age and too much education of team members.[16]
- During recession, avoid cutting budget for all innovative activities. As recession is a cyclic phenomenon, cutting innovative activity budget during this period may severely damage a company's competitive edge.
- Strike a proper balance between creative internal development (i.e., within the company) and externally infused growth through acquisition.

Some of the useful guidelines for managing team members that foster creative work climate are as follows[17]:

- **Focus and gear pressure to goals.** This is basically concerned with establishing definite goals with an approximate time limit for providing the necessary sense of urgency.

- **Allow sufficient time to think.** This is concerned with allowing the team members an adequate amount of time to think with respect to the item in question.
- **Learn how to handle failures.** This is concerned with learning to control the fear of failure by showing tolerance to subordinates' failures.
- **Foster interpersonal contact.** This is concerned with developing effective interpersonal contacts with team members. In this regard, frequent communication, along with freedom for team members to make certain decisions alone, is considered an ideal compromise.
- **Provide quick evaluation and feedback.** This is concerned with providing a quick feedback to team members about their new ideas and at the same time, avoiding premature decisions.
- **Recognize creativity.** This is concerned with recognizing creative individuals through appropriate means (e.g., bringing them to the attention of their peers, superiors, and upper management).
- **Provide a comfortable environment.** This is concerned with providing a comfortable climate to creative individuals through actions such as relieving them, as much as possible, of routine chores and supplying them with appropriate office assistance.
- **Provide continuous feedback.** This is concerned with giving continuous feedback to team members about their newly proposed ideas.
- **Build confidence.** This is concerned with building team members' self-confidence to the maximum. Past experiences indicate that job security, a high degree of job satisfaction, and good pay go a long way in bolstering creative individuals' self-confidence and security.
- **Provide appropriate direction.** This is concerned with providing effective direction to team members through appropriate means.
- **Understand that creativity alone is not enough.** This is concerned with hiring not only those team members who are creative but also others with additional talents.
- **Recognize the need for outside stimuli.** This is concerned with allowing the team members to communicate with peers outside the organization.
- **Balance the need for freedom and the necessity of structure.** This is concerned with balancing the need for freedom and the necessity of structure. More specifically, at times, freedom is needed for creativity; at other times, structure; and sometimes the both.

- **Allow some innocent foibles.** This is concerned with allowing some innocent foibles to the team members. More specifically, avoid denying individuals on a few harmless quirks.
- **Maintain an entrepreneurial climate.** This is basically concerned with creating an entrepreneurial organization within a large bureaucracy.

## 8.6 Tips for Facilitating in a "Cold" Organizational Climate with Respect to Creativity

Some of the useful tips for facilitating in a "cold" organizational climate with respect to creativity are as follows[17]:

- Avoid taking negative remarks personally.
- Take time for the formation of trust between yourself and the group in question.
- Appreciate every small light or signal and reinforce strongly any positive input.
- Aim to be light on aggressive remarks as much as possible.
- Keep up your own energy at an effective level.
- Remain optimistic with respect to changing any negative energy into positive in spite of the difficulties.
- Understand that briefing at the start of a session may be time-consuming under difficult climate. Nonetheless, consider this time as well-spent in order to establish open communication with the group members.
- Avoid overdoing it as the group needs some time to test you and the change it is undergoing.

## 8.7 Workplace Creativity Climate Assessment Checklist

This checklist can be used by managers to assess the friendliness of a given workplace to creativity and innovation. The checklist classifies the assessment areas into seven distinct categories as shown in Fig. 8.3.[18] The response to each item in each category is rated as adequate, strength, or needs improvement. After reviewing the responses to all items in all these categories, the workplace creativity climate can be assessed and appropriate decisions will be made.

The workplace creativity climate assessment-related factors or items belonging to each category of Fig. 8.3 are as follows.[18]

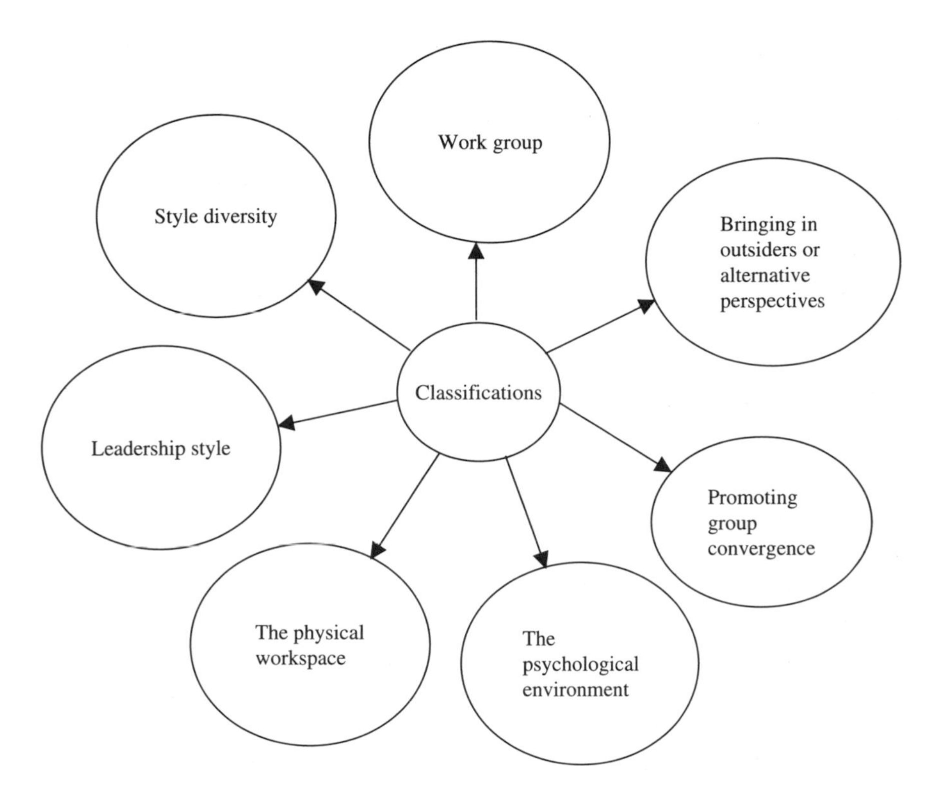


Fig. 8.3. Classifications of workplace creativity climate assessment.

### 8.7.1 *Leadership Style*

- I encourage intellectual conflict among my team members.
- My own preferred style of thinking and working can be rated as it is.
- In my communications with other people, I take into consideration their preferred styles of thinking.
- I have discussed with my team members about their ways of solving problems.
- When a difference of opinions occur among my group members, I help them to determine the cause of their differences.

### 8.7.2 *The Physical Workspace*

- I have made changes to our existing workspace for the purpose of improving communication and creative interaction.

- I encourage my team members to make their workspaces reflect their individuality.
- Our existing workspace incorporates stimulating items such as art, journals, and other objects that are not directly concerned with our business.
- The existing workspace incorporates both areas for quiet reflection and for boisterous interaction.
- I provide my team members a wide variety of traditional and nontraditional tools such as crayons and paper, whiteboards, and e-mails.

### 8.7.3 *Work Group*

- I always keep my eyes open for team members whose thinking styles are different than mine.
- In my group, the majority never ignores the opinions of the minority.
- Members of my team have agreed-upon behavior guidelines for working together and treating each other.
- Our existing work environment is conducive for team members who think differently from the majority.
- I help my team members to develop and agree upon a concise and clear project objective/goal at the beginning of each and every project.
- I have added an individual to my team specifically for bringing a fresh perspective.
- My group members' experiences, skills, and thinking styles are diverse and balanced.

### 8.7.4 *Promoting Group Convergence*

- Project schedules allow sufficient time for group brainstorming sessions and idea discussions.
- At the completion of a project, I provide a way for my team members to rejuvenate and celebrate.
- I encourage my team members to put forward and discuss nonwork-related issues interfering with work.
- Whenever I conduct a debriefing session, I always make sure that all members of the group can attend.
- Whenever my group members are stuck with a problem, I make sure that they get appropriate "down time" or time off to relax and allow their subconscious minds to kick in.

- After the completion of a project, I conduct a debriefing session with my group members for determining specifically what to do differently (or the same) the next time around.

### 8.7.5 *Style Diversity*

- I actively seek or hire individuals having diverse thinking styles and backgrounds.
- We have formally taken diagnostic tests for identifying learning or thinking styles and I have discussed the results of these assessments with my group.
- I understand diverse thinking styles' creative values and I try to include this diversity in my groups.
- My team understands the conflict that creative abrasion can cause understanding of its benefits.

### 8.7.6 *Bringing in Outsiders or Alternative Perspectives*

- My group members have observed customers using our product/service under their own environment.
- My group members have observed various people using competitive products/services.
- I have made appropriate arrangements for speakers from other industries to come talk to my team members.
- My group members visit people outside the division/company for finding different perspectives and ideas.
- We have benchmarked the characteristics and functions of our internal processes, products, or services against an industrial base other than our very own.

### 8.7.7 *The Psychological Environment*

- My group members receive appropriate rewards/recognitions for creative ideas.
- I support my group members taking intelligent risks and do not penalize them when they are unsuccessful.
- There are opportunities for the members of my group for taking on assignments that involve risk and stretch their potential.
- I and my group members assess the risk potential of projects and make appropriate contingency plans or highlight risk management strategies.

- My team members are not panelized for risk taking and experimentation as long as they show learning from the experience.

## Problems

1. Define creativity climate.
2. Discuss the need for having friendly creativity climate.
3. What are the variables that influence peoples' perception of the working climate?
4. What are the important reasons for organizations to foster creativity and innovation?
5. List at least seven attributes of an organization's creative culture.
6. List ten important creative climate dimensions.
7. Discuss important steps for fostering creative environment in companies.
8. List and briefly discuss at least twelve guidelines for managing team members that foster creative work climate.
9. List and discuss important tips for facilitating a "cold" organizational climate with respect to creativity.
10. Discuss seven classification of a workplace creativity climate assessment checklist.

## References

1. Report, The Climate for Creativity, Innovation, and Change, 2003. Available from the Creative Problem Solving Group, Inc., 1325 North Forest Road, Suite 340, Williamsville, New York.
2. Ekvall, G., *Climate in Managing Innovation*, edited by Walker, D. and Henry, J., Sage, London, 1991, pp. 143–150.
3. Kottcamp, E.H. and Rushton, B.M., Improving the Corporate Environment, in *Creativity*, edited by Dale Timpe, A., Facts on File Publications, New York, 1987, pp. 160–165.
4. Plsek, P., Bringing Creativity to the Pursuit of Quality, *Proceedings of the 50th ASQC Annual Quality Congress*, 1996, pp. 99–105.
5. Fisher, A.B., Creating Stockholder Wealth, *Fortune*, Vol. 132, December 1995, pp. 105–116.
6. Deming, W.E., The New Economics, MIT Center for Advanced Engineering Study, Cambridge, Massachusetts, 1993.
7. Utterbeck, J.M., *Mastering the Dynamics of Innovation*, Harvard Business School Press, Boston, 1994.

8. Bower, J.L. and Christensen, C.M., Disruptive Technologies: Catching the Wave, *Harvard Business Review*, Vol. 73, No. 1, 1995, pp. 43–53.
9. Senge, P.M., *The Fifth Discipline: The Art and Practice of the Learning Organization*, Doubleday, New York.
10. Eschenfelder, A.H., Creating an Environment for Creativity, *Research Management*, Vol. 11, No. 4, 1968, pp. 231–241.
11. Farid, F., El-Sharkawy, A.R. and Austin, L.K., Managing for Creativity and Innovation in A/E/C Organizations, *Journal of Management in Engineering*, Vol. 9, No. 4, 1993, pp. 401–409.
12. Meehan, R.H., Programs that Foster Creativity and Innovation, *Personnel*, Vol. 63, No. 2, 1986, pp. 31–35.
13. Amin, M., Development and Leadership of Research Consortia: Lessons Learned and Possible Road Ahead for Continued Innovation, *Proceedings of the IEEE Power Engineering Society Transmission and Distribution Conference*, 2001, pp. 1085–1091.
14. Amabile, T.M., Conti, R., Coom, J., Lazenby and Herron, M., Assessing the Work Environment for Creativity, *Academy of Management Journal*, Vol. 39, No. 5, 1996, pp. 1154–1184.
15. Ensor, J., Cottam, A. and Band, C., Fostering Knowledge Management Through the Creative Work Environment: A Portable Model from the Advertising Industry, *Journal of Information Science*, Vol. 27, No. 3, 2001, pp. 147–155.
16. Zeldman, M.I., How Management Can Develop and Sustain a Creative Environment, in *Creativity*, edited by Dale Timpe, A., Facts on File Publications, New York, 1987, pp. 111–115.
17. Hohn, H. and Verloop, J., Fostering Creativity in Difficult Groups, *International Creativity Network Newsletter*, Vol. 4, No. 2, 1994, pp. 4–5.
18. *Harvard Business Essentials, Managing Creativity and Innovation*, Harvard Business School Publishing Corporation, Boston, Massachusetts, 2003.

## Chapter 9

# Creativity Barriers

### 9.1 Introduction

Creativity is a rather fragile business. It is stimulated by factors such as right environment, co-workers, and self-image, and inhibited by factors such as stresses, limiting beliefs, and work situation.[1] Every person and organization has the potential to make new and better things happen, to bring about new ideas irrespective of their current situations. By becoming aware of what is getting in their way, they can take necessary steps to eradicate obstacles or barriers and start unleashing their creative potential more fully.

Today's fiercely competitive global economy emphasizes more than ever before on companies to remove their existing barriers to creativity. The effectiveness in removing creativity obstacles can be a determining factor in the survival of these companies in today's world economy.

Over the years, many publications discussing various different aspects of the barriers to creativity have appeared.[1–5] These publications appear to be emphasizing, directly or indirectly, the importance of removing creativity-related barriers. This chapter presents various important aspects of creativity barriers.

## 9.2 Reasons for Resistance to Change in Organizations and the Types of Organizations Finding Creativity Most Difficult

Time to time, some organizations may be faced with a significant resistance to change, which may, directly or indirectly, affect creativity and innovation. Some of the possible reasons for the resistance are poor understanding of the purpose of the change, existence of satisfaction with status quo, the change causes anxiety over job security, lack of communication, existence of excessive work pressure and the change is viewed as intensifying it, the individuals affected by the change were not involved in planning for it, lack of confidence in the individual or group initiating the change, fear of failure, existing work-related practices are abruptly changed, the vested interests of the individual or his/her work group are involved, and the change is viewed as providing inadequate rewards or requiring too high a personal cost.[6]

There are many organizations that find creativity most difficult. More specifically, they act as barriers to creativity. They can be categorized under twelve distinct classifications as shown in Fig. 9.1.[7]

## 9.3 Obstacles to Innovation in Large Organizations and Their Overcoming Steps

Past experiences indicated that large organizations with vast resources and sophisticated laboratories, experienced marketing departments, existing

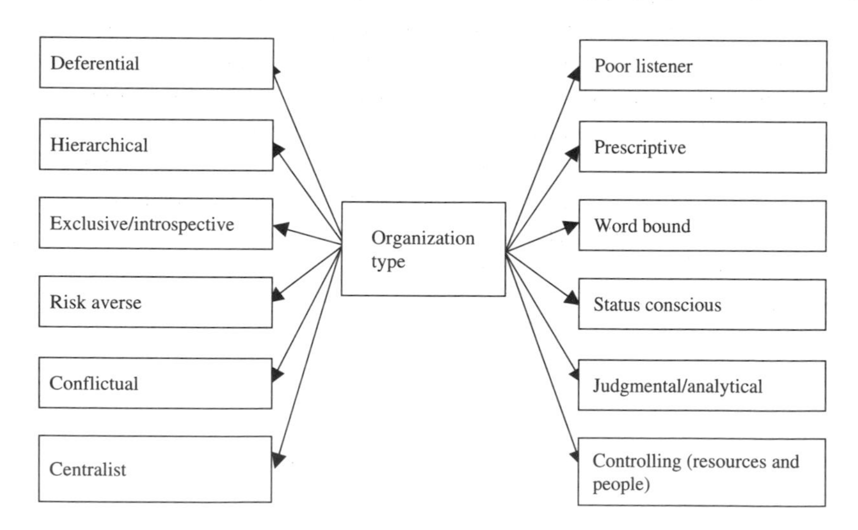


Fig. 9.1. Types of organizations finding creativity most difficult.

channels of distributions, well-known corporate names, etc. are continually beaten to the marketplace by small companies with far fewer resources. The obstacles to innovation in large corporations stem from their shear size with principal objective and strength to manage and maintain an already achieved base or strength. Needless to say, the conservatism and inertia in large corporations give rise to difficulties such as follows[8]:

- **Preclusion of innovation by product/market boundary charters.** It means when development work is in progress within a corporate divisional structure, division business charters sometimes cut off potentially successful innovations because of their incompatibility to the division objectives.
- **Resistance to change because of large size.** More specifically, it can be said that a large organization or corporation is a formalized structure which manages and maintains past successes. Innovation disrupts the stability of the corporate society by interfering with the corporation's vigorous and continuing efforts to be efficient in its current ways and practices. Thus, in large corporations, innovation meets a wall of resistance to change.
- **Tendency to emphasize short-run efficiency.** Large corporations make use of management control system that places emphasis on financial measures. In short, return on investment, short-term profits and the bottom line become the goals and they are measured annually. But as per past experiences, payoffs from innovation are typically five or more years away.[9]
- **Constitution of a "weakest link" constraint on innovation by the separation of power.** Innovation needs both development and marketing of the product. In large corporations usually, both these functions are separated because of the belief that specialization enhances efficiency. Consequently, coordination becomes more difficult and inflexibility occurs. All in all, since vision and enthusiasm are very essential to effective innovation, it is quite difficult for one to sell its innovation-related idea to the other.[10]
- **Current successes may be threatened by innovation**. Often, large organizations are quite reluctant to innovate in areas considered competitive to their existing markets, products, and/or technologies. Needless to say, it is quite unlikely that a company will risk losing its market share and capital investment for the sake of innovation.
- **Development of a short run perspective in managers by the rotation system of managers.** Large corporations train their

managers or leaders by rotating them through the organization. Assignments for fast-track and bright managers are rarely extend beyond two years. As a result, these people fail to see innovations as being within their horizons on any given assignment.

- **The market being followed by the marketing departments rather than leading it.** Inappropriate and narrow application of the marketing concept, time to time, has resulted in large corporations to be driven by the needs of market that are extant and to ignore potential or latent needs of the market.
- **Breeding of conservative subordinates by the corporate hierarchy.** The relationship of boss and subordinate may result in conservatism because subordinates taking risks are exposed to the possibility of bad outcomes. Thus, in large organizations, truly innovative individuals are unlikely to be promoted and rewarded, in comparison to their more conservative counterparts. All in all, it may be concluded that large companies tend to be staffed with more conservative people in regard to innovation.
- **Only getting excited about something big.** Often, large companies or organizations consider a given market opportunity as too small to be interesting. Thus, innovation associated with a market opportunity is overlooked.
- **Belief in growth opportunities through acquisition.** Past experiences indicate that sometimes big corporations believe in terms of achieving growth through acquisition. As this strategy already absorbs successful innovations, it may discourage new developments at home.
- **Politics can result in compromises that weaken the effectiveness of attempts at innovation.** More specifically, the balance of organizational power can result in other difficulties that compromise directly or indirectly innovation.

In large corporations, obstacles to innovation can be overcome by taking actions such as listed in Table 9.1.[8]

## 9.4 Management Barriers to Creativity and Reasons for Prevention of Innovation in Mass-Produced Products

As management plays an important role in the enhancement of creativity, some of its actions could be appropriate while others may be quite detrimental to creativity and innovation. Nonetheless, some of the undesirable

Table 9.1. Actions for overcoming obstacles to innovation in large corporations.

| No. | Action |
|---|---|
| 1 | Establish goals for innovative achievement |
| 2 | Reward successful innovation |
| 3 | Encourage managers to take a long-term perspective |
| 4 | Carefully screen customers to identify new ideas |
| 5 | Make special unrestricted funds available for exploring innovative ideas without upper-level approval |
| 6 | Accept failures as part of the game |
| 7 | Encourage engineers, scientists, etc., to meet customers |

management barriers to creativity and innovation are as follows[11,12]:

- Too rigid a line of authority and nonexistence of long-range goals or objectives.
- Poor communication between management and workers and the tendency of the management to tell professionals or others rigidly what to do and how to do it.
- Failure to recognize creative people.
- Failure to reward creative people.
- Too much emphasis on quick use of ideas.
- Reluctance to take chances and making frequent changes to important decisions.
- Poor or incorrect handling of credit for new ideas.
- Expression of a rather negative attitude towards all new ideas.

Over the years, various professionals have observed many reasons for the delay or prevention of innovations in mass-produced products. Some of the important reasons are shown in Fig. 9.2.[13]

## 9.5 Ways for Managers to Kill Creativity and Ways Used by Technical Managers to Block Creative Ideas

There are managers who do not want to support creative ideas to blossom in their organizations. These managers can kill creative ideas through the following twelve ways[14,15]:

- **Hold many meetings.** This will certainly kill time and the interest of other people.

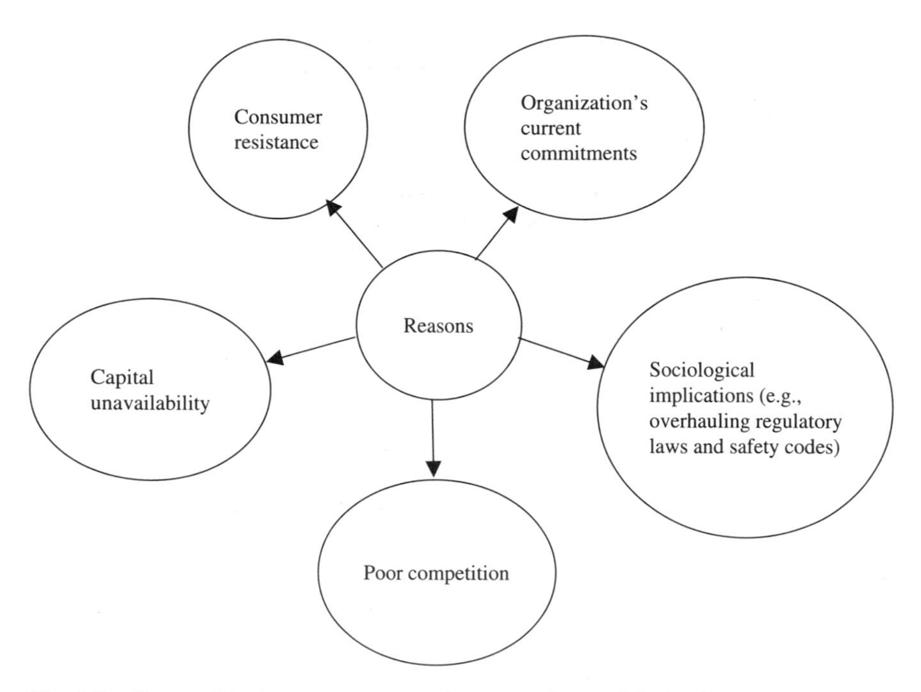


Fig. 9.2. Some of the important reasons for preventing or delaying innovation in mass-produced products.

- **Patiently wait for the result of market surveys.** They take time. It could be too much!
- **Patiently wait for detailed analysis.** In this case, logic appears to be on the side of managers.
- **Drag your feet.** Who can forcefully argue with painstaking managers or leaders?
- **Do not follow up in an effective manner.** It means drop the idea among concerned individuals and then let them worry about it.
- **Inflate the projected cost estimates.** This will guarantee managers heroes' role for saving the company money when the idea under consideration is vetoed.
- **Strictly follow protocol.** It means red tape will be around for a long time after the dead idea.
- **Lack a sense of urgency.** It basically means that organization's business has survived and will survive for a long time to come, so why rush?
- **Agree to the idea, but do not act upon it.** This type of diplomatic action leaves almost everyone pleased for a while.

Table 9.2. Ways used by some technical managers to block creative ideas.

| No. | Way |
|---|---|
| 1 | It is too academic |
| 2 | It will not work |
| 3 | It is against company policy |
| 4 | It requires more study |
| 5 | It is too old fashioned |
| 6 | It will be too difficult to implement |
| 7 | Let's form a committee |
| 8 | We do not have the right manpower |
| 9 | We are too big |
| 10 | We have tried it before |
| 11 | Union will scream |
| 12 | The top management will not buy it |
| 13 | Let's conduct a survey first |
| 14 | We have too many ongoing projects at the moment |
| 15 | It will be too difficult to administer |
| 16 | It will cost too much |
| 17 | It is not our problem |
| 18 | We are too small |
| 19 | Our customers may not like it |
| 20 | Did some one else tried it? |
| 21 | It is not in your job description |
| 22 | It is too radical |

- **Show concerns about the budget.** It will basically say why invest when things are going well?
- **Steer the idea under consideration into channels.**
- **Encourage the "Not Invented Here" type syndrome.**

Over the years, some technical managers have used various ways to block creative ideas. Some of these ways are presented in Table 9.2.[15]

## 9.6 Stumbling Blocks and Building Blocks to Creativity

There are many organizations that create mental blocks to creativity without even noticing it. Some of these mental or stumbling blocks along with their corresponding building blocks to creativity in parentheses, are lack of flexibility (flexibility), discomfort with taking risks (intelligent risk taking), resource myopia, i.e., nearsightedness (resourcefulness), seeing play as only frivolous (playfulness), intolerance (tolerance), following the rules, i.e., too closely, to often (ability to think outside the rules), fear of failure (ability to accept failure and learn from it), giving up too soon (persistence), political problems and turf battles (collaboration, focus on mutual gain), focusing on just the right answer (focus on exploring possibilities), thinking of not having creative talent (recognizing creative potential in self), being judgmental, critical (being accepting), worrying too much about what others will think (having an inner focus), avoiding ambiguity (tolerance for ambiguity), lack of openness to ideas (receptivity to ideas), and difficulty in hearing another perspective or opinion (active listening, acceptance of differences).[16]

A careful examination of these stumbling blocks can be useful to assess if blocks such as these are a factor in one's organization. Similarly, the building blocks to creativity such as these can be quite useful to nourish creativity in one's organization.

## 9.7 Types of Barriers to an Individual's Creative Thinking and Suggestions for Overcoming Them

There are many barriers that may inhibit creativity of any individual or professional. They may be grouped under four classifications as shown in Fig. 9.3.[12,17–19]

The emotional barriers are self-imposed and are probably the most serious barriers of all. Some examples of these barriers are as follows[20]:

- Fear
- Self-satisfaction

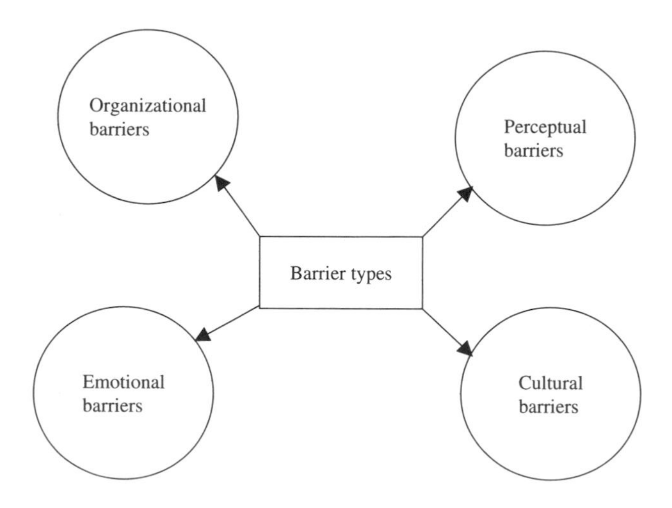


Fig. 9.3. Types of barriers to creative thinking.

- Greed
- Poor self-confidence
- Resistance to change
- Unwillingness to accept help from others
- Over cooperation
- Hate
- Desire for security

The perceptual barriers are the result of the mind's tendency to short circuit. The main reason for their occurrence is the failure to recognize each situation's components as individual elements. For example, when a person becomes over-acquainted with some items, then he or she tends to overlook the background of such items in detail.

The cultural barriers are imposed by the society and environment on an individual or a person. Some of their examples are competitiveness, cooperation, and conformity. The organizational barriers are facilities-oriented and set primarily by the management.

Some of the suggestions for overcoming each of the above four types of barriers are as follows[17,21,22]:

**Emotional barriers**

- Develop new ideas at pace suitable to you, regardless of the pressure or any other factor.
- Always champion your ideas even in the face of obvious rejection.
- Make a serious personal commitment to your creative ideas.
- Avoid worrying about failure if you are seeking success seriously.

**Perceptual barriers**

- Break down the problem into elements and rearrange the elements.
- Fully challenge your assumptions.
- Determine an appropriate analogy.
- Turn the problem under consideration upside down.
- Try a wild approach.

**Cultural barriers**

- Unleash your personal feelings.
- Fantasize your Disney World.
- Unleash your personal preferences.

### Organizational barriers

- Identify important problems.
- Focus on these important problems.
- Reorganize your workplace or work area.
- Notify your colleagues/coworkers and others of your earnest desire for privacy.
- Develop and post your schedule of no-interruption time periods.

## 9.8 Creativity Inhibitors an Engineer May Encounter While Inquiring into and Solving the Problem

There are a great many creativity inhibitors an engineer may encounter while inquiring into and solving the problem. A comprehensive list of such inhibitors is presented in Ref. 23. Some of these inhibitors are as follows[23]:

- Conformity
- Timidity
- Excessive self criticism
- Frustration
- Fear of failure
- Fear of loosing job
- Fear of new ideas
- Fear of questioning
- Fear of making mistakes
- Habits of familiarity with the problem
- Rebellion against management
- Rebellion against colleagues
- Rebellion against supporting help
- Satisfaction with status quo
- Deliberately avoiding the problem
- Worry
- Inability to relax and free ideas
- Loss of perspective of importance problem
- Loss of patience and persistence
- Loss of curiosity
- Loss of financial support
- Excessive pressure to create
- Excessive perfectionism

- Excessive analysis at wrong time
- Excessive supervision
- Excessive emphasis on cost at wrong time
- Excessive marketing problems
- Excessive legal difficulties
- Insensitivity to recent advances
- Inadequate knowledge/information
- Incorrect assumptions
- Overspecialization
- "Giving up" after first unsuccessful experiments
- Lack of correct problem definition
- Lack of encouragement
- Lack of proper materials
- Lack of time allocation to do creative work
- Lack of an action plan to create
- Wrong decisions
- Lack of adequate recognition/rewards for creating
- Lack of commitment to solve problem
- Limitation on experimental facilities
- Unnecessary secrecy
- Unrealistic time schedules
- Frequent changes to key decisions
- Dishonest promoters
- Militaristic obedience to "authorities"
- Resistance to change
- Spirit of dogmatism

## 9.9 Barriers to Creativity in Textile Industry

Creativity is considered an important factor in textile industry. A study conducted in the United Kingdom (UK) has identified many barriers to creativity in textile industry in UK.[24] Some of those barriers are as follows[24]:

- Practice of a traditional management style and a fixed hierarchical organizational structure by most companies.
- The tendency of companies to minimize risk through review and control.
- Poor or no feedback system employed by companies to check if employees fully understood and accepted the company vision.
- Tendency of employees to regard criticism as nonconstructive.

- Existence of no specific facilities in companies for encouraging creativity.
- A practice of copying competitors' new product ideas in most companies.
- Existence of a strained relationship between research and development (R & D) and marketing departments.
- A rare use of teams for new product idea generation.
- Insufficient knowledge of employees concerning the usage of their products by customers.
- A rare existence of mechanisms for collecting ideas.

## Problems

1. List principal reasons for resistance to change in organizations.
2. List at least ten types of organizations that find creativity most difficult.
3. Discuss obstacles to innovation in large organizations and their overcoming steps.
4. Discuss management barriers to creativity.
5. What are the important reasons for the prevention of innovation in mass-produced products?
6. List at least twelve ways in which managers can kill creativity in their organizations.
7. Discuss four types of barriers to an individual's creative thinking.
8. List at least forty creativity inhibitors an engineer may encounter while inquiring into and solving the problem.
9. Discuss barriers to creativity encountered in the textile industry.
10. Write an essay on creativity barriers in engineering organizations.

## References

1. Selwyn, P., Twenty Three Blocks to Creativity. Available from the Innovation Network, 451 E. 58th Ave, Suite 4625 and Box 468, Denver, Colorado, USA.
2. Amabile, T., How to Kill Creativity, *Harvard Business Review*, Vol. 76, 1998, pp. 76–87.
3. Dunn, D.T., The Rise and Fall of the Venture Groups, *Business Horizons*, October 1973, pp. 32–41.
4. Schon, D.A., *Technology and Change*, Delacorte Press, New York, 1967.
5. "How Four Companies Spawn New Products" by Encouraging Risks, *The Wall Street Journal*, 18 September 1980.
6. Azani, H. and Khorramshahgol, R., The Impact of Automation on Engineers' Creativity and Innovation and Its Implications for Reducing Resistance to Change, *Computers in Industry*, Vol. 16, 1991, pp. 377–383.

7. Bichard, M., Creativity, Leadership, and Change, *Public Money and Management*, April–June, 2000, pp. 41–46.
8. McIntyre, S.H., Obstacles to Corporate Innovation, in *Creativity*, edited by Dale Timpe, A., Facts on File Publications, New York, 1987, pp. 116–123.
9. Fast, N.D., New Venture Departments: Organizing for Innovation, *Industrial Marketing Management*, Vol. 20, 1978, pp. 77–88.
10. Schon, D.A., *Technology and Change*, Delacorte Press, New York, 1967.
11. Shannon, R.E., *Engineering Management*, John Wiley and Sons, New York, 1980.
12. Dhillon, B.S., *Engineering and Technology Management*, Artech House, Inc., Boston, 2002.
13. Harrisburger, L., *Engineermanship: A Philosophy of Design*, Wadsworth Publishing Company, Belmont, California, 1966.
14. Fraenkel, S.J., How Not to Succeed as an R & D Manager, *Research Management*, May 1980, pp. 35–37.
15. Badawy, M.K., How to Prevent Creativity Mismanagement in *Creativity*, edited by Dale Timpe, A., Facts on File Publications, New York, 1987, pp. 176–188.
16. *Harvard Business Essentials: Managing Creativity and Innovation*, Harvard Business School Publishing Corporation, Boston, 2003.
17. Farid, F., El-Sharkawy, R. and Austin, L.K., Managing for Creativity and Innovation in A/E/C Organizations, *Journal of Management in Engineering*, Vol. 9, No. 4, 1993, pp. 401–409.
18. Karger, D.W. and Murdick, R.G., *Managing Engineering and Research*, Industrial Press, New York, 1969.
19. Adams, J.L., *Conceptual Blockbusting: A Guide to Better Ideas*, Addison-Wesley, Reading, Massachusetts, 1986.
20. Mason, J.G., *How to be a More Creative Executive*, McGraw Hill Book Company, New York, 1960.
21. Feinberg, M.R., *Effective Psychology for Managers*, Prentice Hall, Inc., Englewood Cliffs, New Jersey, 1965.
22. Morgan, J.S., *Improving Your Creativity on the Job*, American Management Association, New York, 1968.
23. Bailey, R.L., *Disciplined Creativity for Engineers*, Ann Arbor Science Publishers, Ann Arbor, Michigan, 1978.
24. McAdam, R. and McClelland, J., Sources of New Product Ideas and Creativity Practices in the UK Textile Industry, *Technovation*, Vol. 22, 2002, pp. 113–121.

## Chapter 10

# Creativity in Quality Management, Software Development Process, Rail Transit Stations, and Specific Organizations

## 10.1 Introduction

Although the term creativity has been around for a long time and is being used in many areas, there is no widely accepted definition of creativity. In fact, as per Ref. 1, there are over one hundred definitions of creativity (e.g., characters of a person, characters of one product or outcome, or processes through which creative people can produce creative results.[2,3])

Nonetheless, in our present dynamic world, the importance of creativity and innovation can hardly be overestimated because the competitive advantage of today is becoming a common practice of tomorrow.[4] For organizations wanting to be at the forefront of a business, continuous creativity in their products and processes appears to be a good strategy. This requires a careful planning and appropriate actions from management with respect to creativity.

Needless to say, today, a careful consideration to creativity is being given in various industrial sectors. This chapter presents various different aspects of creativity in quality management, software development process, rail transit stations, and specific organizations.

## 10.2 Creativity in Quality Management

Creativity and quality management have much in common as both are fundamentally related to the organization's success. As customer needs drive both the pursuit of quality and innovation, a focus on quality and creativity/innovation is essential for beating the competition. Creativity tools can help to solve quality-related problems and redesign processes to improve customer satisfaction and reduce waste. Needless to say, as per W. E. Deming, joy in work helps to achieve higher levels of quality and creative thinking is one way of building that joy.[5,6]

### 10.2.1 *Customer Needs Analysis*

As the concept of understanding and exceeding the needs of customers is central to quality management, companies spend millions of dollars each year to collect information about customers through focus groups, surveys, and market research. The notion of customer needs analysis is natural for the directed creativity and the quality management. Eight ways for being more creative in customer needs analysis are shown in Fig. 10.1.[6]

Way I (i.e., who is the customer?) is concerned with making a list of no obvious customers or people who could be the beneficiaries of the product, service, or the process under consideration as well as determining the desire of these people. Way II (i.e., how could we use it?) is concerned with taking parts of existing products, services, and process to hunt deliberately for customer needs they might satisfy. Here, the main goal is to surprise customers with new uses for familiar items/products.

Way III (i.e., dimensions of quality analysis) is concerned with separating quality dimensions into two categories (i.e., developed and nondeveloped) and then exploring with customers the better ways and means for delivering the poorly developed dimensions. Way IV (i.e., bring prototypes

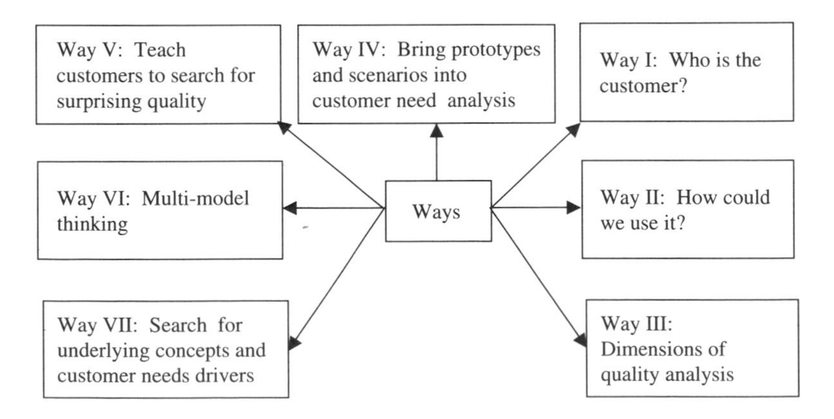


Fig. 10.1. Ways for being more creative in customer needs analysis.

and scenarios into customer need anlaysis) is basically concerned with making the ideas more concrete with prototypes or detailed scenarios and then analyzing reactions of customers.

Way V (i.e., teach customers to search for surprising quality) is concerned with teaching customers directed creativity and then allowing them to use their unique mental valleys for generating creative ideas for you. Way VI (i.e., multimodel thinking) is concerned with using de Bono's Six Thinking Hats as a framework for asking customers to generate lists of positive and negative feelings and creative ideas, and then utilizing all these to creative provocations for future idea generation sessions. Way VII (i.e., search for underlying concepts and customer needs drivers) is concerned with taking a customer complaint, behavior, or comment and then asking "why do customers care about this?" Here, the main objective is to highlight purpose concepts that can be utilized as fixed points for alternatives.

## 10.3 Creativity in Software Development Process

Today, the software development process requires more creative or innovative thinking than ever before. In the modern age of the Internet, the challenge appears to have shifted considerably from writing the code to identifying and evaluating innovative ideas, processes and applications. Creating software is one of the most creative activities that people undertake and the major limitation in software is the human imagination, and the limitations on that are essentially all self-imposed. Needless to say, the application of creativity makes it possible to create truly great software.

### 10.3.1 *Creativity Methods in the Software Development Process*

Various creativity methods are used in the software development process. Three important reasons for their usage are establishing open communications among all involved parties, focusing on cause–effect relationships, and redefining the problem.[7] Useful creativity techniques for each stage of the software development process (i.e., requirement definition, design, coding, testing, or evaluation) are presented below.

#### 10.3.1.1 *Requirements Definition*

The requirements definition stage is probably the most crucial to the entire project and its main/overall goal is to identify user requirements.[8]

Two promising creativity techniques for use during this stage are Wishful Thinking[9] and Boundary Examinations.[10]

Wishful Thinking exercises can be used by permitting a group of individuals to assume that anything and everything is feasible in addressing a problem statement. Subsequently, the group members find that even though certain things were impossible, an effective substitute is found. When an effective and feasible alternative to the impossible task is found, the problem is reinstated. This method is described in detail in Ref. 9.

Boundary examination exercises focus on problem assumptions. Group members draft an initial problem statement and then examine it from any hidden assumptions as well as they identify implications of these assumptions. Consequently, the group revises the problem statement. This method is described in detail in Ref. 10.

#### 10.3.1.2 *Design*

The main goal of the design stage is to determine the proposed system's structure. New ideas concerning the proposed system structure are generated and tested through others' reviews. Two creative problem-solving methods considered useful for the idea generation part of this stage are Input–Output[11] and Goal Orientation.[9]

The Input–Output technique carries out various functions that are included in what have been often known as "structured" methods in system development.[12,13] Such methods are quite useful for isolating where objects such as data items are transformed and for coordinating a project with many such transformations. This technique is described in detail in Ref. 11. Goal Orientation exercises can be used by having the group highlight the fundamental problem and then specify constraints, requirements, and obstacles associated with the problem. There could be questions that may suggest redefinition of the design problem. This method is described in detail in Ref. 9.

#### 10.3.1.3 *Coding*

This stage is concerned with constructing the system in great detail. More specifically, it requires programmers to generate ideas about how the design might be accomplished physically. Two creative problem-solving methods considered useful for this stage are Reverse Brainstorming[11] and Analogies.[14]

Reverse Brainstorming is often used in software-related departments where during code inspections, a group of individuals criticize already written program statements. In places where Reverse Brainstorming is followed more closely, a meeting is held to review all potential methods for coding major modules. A list of ideas is generated by the group members for each reviewed module and then each idea is discussed critically. The method resulting in least serious criticisms is adopted. Reverse Brainstorming is described in detail in Ref. 11.

The Analogies method is usually used, in part, in software-related organizations that use code libraries. This method requires that the problem under consideration be stated clearly, a specific situation be related to the problem in the form of analogy, and examining the relationship of the analogy to the basic/original problem. This method is described in detail in Ref. 14.

#### 10.3.1.4 *Testing*

This stage is basically concerned with developing test plans and generating test data. Software testing is unusual in the sense because here one deliberately attempts to introduce chaos (i.e., erroneous data) into the system, while creativity methods strive to determine order out of chaos. Nonetheless, two promising creative problem-solving methods for the testing stage are Nonlogical Stimuli and Simulation.[9]

The Nonlogical Stimuli technique is a loosely-structured approach carried out by individuals. In this method, the test data are generated by continually attempting to introduce random elements into the system. Nonetheless, the method is described in detail in Ref. 9.

The Simulation approach is another promising creative problem-solving method for the testing stage. In this method, group members generate as many ideas as possible with respect to how errors might occur and then refine each of these ideas into one or more test procedures. The Simulation method is described in detail in Ref. 9.

#### 10.3.1.5 *Evaluation*

This is the final stage of the software development process and is concerned with generation and selection of criteria for evaluation, data collection planning and execution, and communication of evaluation results/conclusions to others. Two most promising creative problem-solving methods for this stage are the Nominal Group Technique[15,16] and Morphological Analysis.[17,18]

The Nominal Group Technique is useful to avoid problems of dominance by certain people, haste, and premature closure. Furthermore, it can be an effective tool to develop a procedure for system evaluation. In this approach, individuals can generate ideas of evaluation processes and criteria separately in writing, "round-robin" recording of ideas, clarifying discussion, etc. Additional information on this method is given in Refs. 15 and 16.

Morphological Analysis calls for dividing problems under consideration into two or more dimensions and then investigating the existing relationships between these dimensions. For example, in the evaluation of an information systems project, an individual or a group of individuals can concentrate on certain aspects of success (e.g., satisfaction, availability, and efficiency) and how they relate to constituencies such as analysts, coders, managers, and customers. This method is described in detail in Refs. 17 and 18.

## 10.4 Creativity in Rail Transit Stations

Just like in any other areas of engineering applications, creativity is considered an important element in rail transit stations. The design of many urban rail transit stations clearly demonstrates the integrated creative efforts of artists, architects, and engineers. The fusion of creativity in a number of rail transit stations/systems is discussed below, separately.[19]

### 10.4.1 *Washington's Union Station*

The Union Station of Washington, D.C. is an important example of the beginning of the fusion of creativity in rail transit stations. The station was conceived at the beginning of the twentieth century and developed during the heyday of railroading. The use of subtle color patterns, choice of materials, poetic statements carved directly into the granite façade, sophistication of detailing, and direct incorporation of large sculptural objects clearly demonstrate the reaching of a pinnacle of fusion rarely achieved in public building structures since the days of construction of great religious monuments during the Renaissance in Europe. Additional information of the fusion creativity in this station is available in Ref. 19.

### 10.4.2 *Stockholm Subway/Underground*

This subway/underground is another important example of the fusion of creativity in rail transit stations. This underground rail system came into existence in the late 1940s and early 1950s and with the establishment of the Stockholm Transport Art Advisory Council, the real achievements in integrating art with architecture and engineering came around during the early part of the 1970s.[20] The complexity, variety, quality of imagery, and execution within this totally subterranean environment appear to have reached the highest level of fusion. In particular, the Radhuset Station platform clearly demonstrates that something well beyond pragmatic considerations of flow, safety, comfort, orientation, and function has been achieved in an effective manner in the cavernous stations. Additional information on the fusion of creativity in Stockholm Underground is available in Ref. 20.

### 10.4.3 *Moscow Underground System*

This system was conceived in the late 1920s and the early 1930s, and presents another shining example of an effective achievement in integration of engineering, art, and architecture in underground environment. The underground stations are richly ornamented and dramatized through the use of color, lighting, and decorative art, and have become museums to display the creative talent in Russian Society.[21] Additional information on the fusion of creativity in Moscow Subway System Stations is available in Ref. 21.

### 10.4.4 *Atlanta Rapid Transit System*

This system, often known simply as MARTA (i.e., Metropolitan Atlanta Rapid Transit Authority), represents the beginning to recognize the role of artists as part of the station design team. In many MARTA stations, commissioned artists produced objects for installation in predetermined spaces, with the imagery, in many instances reflecting neighborhood ethnic, historic, social, or cultural factors.[22] In particular, Peachtree Center is probably the most dramatic station in the Atlanta system. In this case, both engineers and architects recognized the potential for creating a magnificent station from the exposed granite cavern resulting from blasting through the rock.[19] Needless to say, here engineering technology itself has produced a rich environment with strong colors and textures; although all the essential elements of lighting, ductwork, and rock bolts are integrated with care into the stone cavern itself.

Additional information on the fusion of creativity in the MARTA system stations is available in Refs. 19 and 22.

### 10.4.5 *Washington's Metro System*

The Metro system of Washington, D.C., represents the epitome of engineering and architecture as art in its barrel vault forms, dramatic lighting, subtle use of color, and material simplicity. The structural engineering need to span a fairly broad space accommodating platforms and tracks without the use of any column has created the dramatic shallow-arched forms; resulting in an unparallel consistency and uniformity by any other transit system in the country.

Additional information on the fusion of creativity in this system is available in Ref. 19.

### 10.4.6 *Seattle Metro System*

This underground system has a number of stations that represent a true fusion of creativity. A number of artists worked alongside with the design teams for each station; thus were able to involve directly in virtually all decisions concerning items such as materials, colors, textures, lighting fixtures, and handrails. All in all, this project is considered as a testament to the determination of all involved parties to achieve spectacular end results.

Additional information on the fusion of creativity in Seattle Metro System is available in Ref. 23.

## 10.5 Creativity in Specific Organizations

Successful organizations are very effective in creating a successful advantage for their products or services through creativity and innovation. They are creative and innovative not by accident, but by effectively managing their human resources for creating and marketing new products and services. Needless to say, as people are the most vital resource of a successful company, the successful innovation organizations have learned quite effectively how to motivate, manage, and reward their employees.

Various aspects of creativity in a number of specific organizations are discussed below, separately.

### 10.5.1 *Creativity in Hewlett-Packard*

This company has established a framework for integrating the notions of change, cross-functional teams, and creativity.[24] On the basis of this

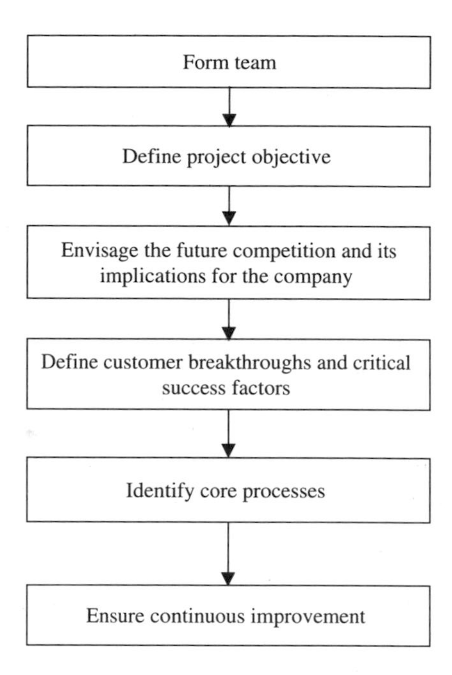


Fig. 10.2. Hewlett-Packard creative team process steps/stages.

framework, Hewlett-Packard has been very successful in improving the performance of its cross-functional teams with respect to product and operational excellence. The organization uses this framework both internally and externally at the time of introducing radical change projects. The framework (i.e., the creative team process) is based on the experiences gained from many change projects. It is carried out in the form of workshops at the initial stage of a change project (e.g., cycle time reduction). The creative team process is composed of six steps (i.e., stages) as shown in Fig. 10.2.[24]

Each of the steps/stages is discussed in detail in Ref. 24.

### 10.5.2 *Creativity in Motorola*

Motorola has formed several in-house teams for identifying new product ventures.[25] A team is typically made up of five to six individuals belonging to areas such as marketing, engineering, research and development, manufacturing, sales, and finance. These individuals are expected to display tolerance to ambiguity, eagerness for new experiences, and openness to new ideas and thoughts.[26]

In addition to these teams, Motorola, also, makes use of various interdepartmental functional teams for developing new products. These teams get involved in the new product development process right from the start of the project. Needless to say, Motorola's approach with respect to newly-developed product has resulted in many benefits including cost-effectiveness in manufacturing, an effective meeting of customer satisfaction criteria, and a potential for bringing large profits to the company.

### 10.5.3 *Creativity in the 3M Company and in Apple Computer Organization*

The 3M Company uses venture teams in developing new products. Teams are composed of individuals with various skills and disciplines including the employees who "Champions" the ideas of a new product. At one point in time, the team members commit themselves 100% to one specific product venture and put all other projects on the back burner.[27] Additional information on the topic is available in Refs. 25 and 27.

In the new product development process, the Apple Computer organization also forms venture teams by balancing team members' skills and personalities. During the team formations, the organization gives particular attention to individuals with excellent technical knowledge and broad mind in understanding others' opinions and goals, in addition to exercising good people skills.[28]

## Problems

1. Discuss creativity in quality management.
2. What is customer needs analysis?
3. What are the ways for being more creative in customer needs analysis?
4. Discuss creativity in the software development process.
5. Discuss useful creativity methods for use in software requirements definition, design, and coding phases.
6. Discuss the fusion of creativity in the following rail transit stations/systems:
   - Washington's Union Station.
   - Moscow Underground System.
   - Atlanta Rapid Transit System.
7. Discuss creativity in the Hewlett-Packard company.

8. Discuss creativity in Motorola with respect to new product ventures.
9. Discuss useful creativity methods for use in software testing and evaluation phases.
10. Discuss creativity in the 3M company and in Apple Computer organization with respect to new product development.

## References


1. Gu, M. and Tong, X., Towards Hypotheses on Creativity in Software Development, available from the Department of Computer and Information Science, Norwegian University of Science and Technology, Trondheim, Norway.
2. Gough, H.G., A Creative Personality Scale for the Active Checklist, *Journal of Personality and Social Psychology*, Vol. 37, 1979, pp. 1398–1405.
3. Amabile, T.M., A Model of Creativity and Innovation in Organizations, *Research in Organizational Behaviour*, Vol. 10, 1988, pp. 123–167.
4. Nijhof, A., Krabbendam, K. and Looise, J.C., Innovation Through Exemptions: Building Upon the Existing Creativity of Employees, *Technovation*, Vol. 22, 2002, pp. 675–683.
5. Plesk, P.E., Incorporating the Tools of Creativity into Quality Management, *IEEE Engineering Management Review*, Fall 1998, pp. 61–68.
6. Plesk, P.E., *Creativity, Innovation, and Quality*, ASQ Quality Press, Milwaukee, Wisconsin, 1997.
7. Galletta, D.F., Sampler, J.L. and Teng, J.T.C., Strategies for Integrating Creativity Principles into the System Development Process, *Proceedings of the Hawaii International Conference on System Science*, 1992, pp. 268–276.
8. McKeen, J.D., Success Development Strategies for Business Application Systems, *MIS Quarterly*, September 1983, pp. 47–65.
9. Rickards, T., *Problem Solving Through Creative Analysis*, Gower Press, Essex, UK, 1974.
10. Arnold, J.E., The Creative Engineer, *Yale Scientific Magazine*, March 1956, pp. 12–23.
11. Whitting, C.S., *Creative Thinking*, Reinhold Book Company, New York, 1958.
12. Jackson, M.A., *System Development*, Prentice Hall, Inc., Englewood Cliffs, New Jersey, 1983.
13. Yourdon, E. and Constantine, L.L., *Structured Design*, Prentice Hall, Inc., Englewood Cliffs, New Jersey, 1979.
14. De Bono, E., *Lateral Thinking: Creativity Step by Step*, Harper and Row, New York, 1970.
15. Delbecq, A. and Van de Ven, A., A Group Process Model for Problem Identification and Program Planning, *Journal of Applied Behavioral Science*, Vol. 7, 1971, pp. 466–492.
16. Delbecq, A., Van de Ven, A. and Gustafson, D., *Group Techniques for Program Planning*, Scott, Foresman, Glenview, Illinois, 1975.

17. Allen, M.S., *Morphological Creativity*, Prentice Hall, Inc., Englewood Cliffs, New Jersey, 1962.
18. Zwicky, F., *Discovery, Invention, Research Through the Morphological Approach*, MacMillan, New York, 1969.
19. Kivett, H.A., Fusion of Creativity in Rail Transit Stations: A Retrospective and Critique, *Transportation Research Record*, No. 1549, 1996, pp. 75–78.
20. Thompson, L., Art Goes Underground, Art in the Stockholm Metro, Stockholm Transport Art Advisory Committee, Stockholm, Sweden, 1989.
21. Berezin, V., *Moscow Metro Photo Guide*, Planeta Publishers, Moscow, Russia, 1989.
22. MARTA Station Art Book, Metropolitan Atlanta Rapid Transit Authority, Atlanta, Georgia, 1989.
23. Kartiganer, L., In Collaboration: Artists on the Metro Design Team, Municipality of Metropolitan Seattle, Seattle, Washington, 1992.
24. Feurer, R., Chaharbaghi, K. and Wargin, J., Developing Creative Teams for Operational Excellence, *International Journal of Operations and Production Management*, Vol. 16, No. 1, 1996, pp. 5–18.
25. Gupta, A.K. Singhal, A., Managing Human Resources for Innovation and Creativity, *Research Technology Management*, Vol. 36, No. 3, 1993, pp. 41–48.
26. Kapp, S., Lawyer Turned Marketing Crusader, *Business Marketing*, July 1987, pp. 12–16.
27. Masters of Innovation, *Business Week*, 10 April 1989, pp. 58–63.
28. Apple Computer Tries to Achieve Stability But Remain Creative, *Wall Street Journal*, 16 July 1987, p. 1.

# Chapter 11

# Creativity Testing, Recording and Patents

## 11.1 Introduction

The identification of individuals with special creative talents has been a challenging issue to many creativity researchers because of a heavy demand from various sectors of economy. In fact, over the past fifty years there has been lot of ongoing efforts to identify creative individuals by determining their personality types. As the result of this effort, many creativity tests have been developed.[1–3]

For future use and in the use of ongoing creativity efforts, a good documentation of all creative thoughts is essential. Fortunately, there are many ways and means for keeping records of creativity.[4] However, their effectiveness depends upon the degree of care given in selecting a particular method for a specific application.

A patent may simply be described as an exclusive right of its owner to exclude other people from using, making, or selling the creative invention as outlined in the claims of the patent for the stated time period. As per Franklin D. Roosevelt, "Patents are the key to our technology; technology is the key to production".[5] Needless to say, almost every engineering professional is affected by the patent system.

This chapter presents various important aspects of creativity testing, recording and patents.

## 11.2 Creativity Testing

There are various ways and means used in creativity testing. A typical creativity test attempts to determine whether or not the person taking the test has capabilities, the developer(s) of the test identify with creativity and innovation. Some different aspects of the creativity testing are presented below.

### 11.2.1 *Creativity Tests*

Creativity tests may be divided into four main categories as shown in Fig. 11.1.[1–3] These are divergent thinking, self assessments, convergent thinking, and artistic assessments.

Divergent thinking is concerned with one's ability to consciously produce new ideas that result in many possible solutions for a given problem. In turn, the solutions are evaluated with respect to four factors: originality, fluency, flexibility, and elaboration.

Artistic assessments are concerned with the evolutions of an artistic product such as a painting, a musical composition, or a story. Usually, the evaluations are performed by two or more judges and their judgments or evaluations must be in close agreement on the product creativity.

Convergent thinking is concerned with one's ability to correctly hone in the single correct solution to a given problem. In turn, the solution is assessed either right or wrong. Self assessments are persons' responses to the amount of creativity (i.e., the persons feel they exhibit).

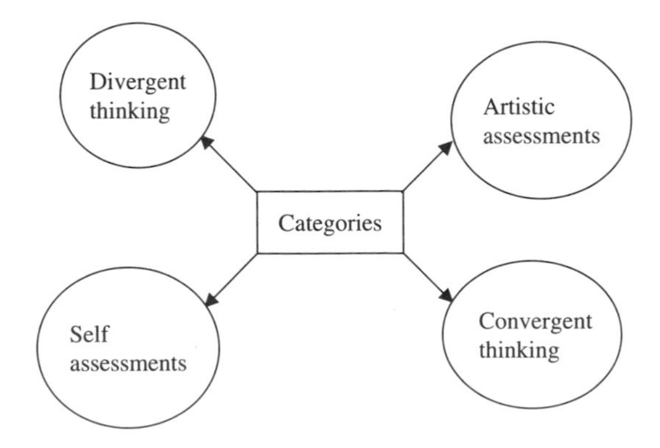


Fig. 11.1. Main categories of creative tests.

Two specific creativity tests pertaining to the divergent thinking category are as follows:

- **Wallas and Kogan Test.**[6] In this creativity test, individuals are asked to come up with many possible items that contain a specific component (e.g., name things with wheels). The test can be administered by any one and its scoring is based upon four factors: originality, flexibility, elaboration, and fluency.
- **Guilford's Alternative uses Task Test.**[7] In this test, individuals are asked to list as many as possible uses for a common house hold item (e.g., name all the uses for a brick). Its scoring is based upon four factors: flexibility, originality, fluency, and elaboration.

## 11.2.2 *Specific Creativity Tests*

This section presents two useful specific creativity tests.

### 11.2.2.1 *Test I*

This test is designed to assess the degree of creativity a person possesses. The test requires the individual to answer true or false to a list of ten items presented in Table 11.1.[8]

Answers to Table 11.1 items 1–10 are false, true, false, true, false, true, true, false, false, and false, respectively. Explanations of these answers are

Table 11.1. A list of items (i.e., quiz) for testing the creative ability of a person.

| No. | Item |
|---|---|
| 1 | It is always easier to find solution to a problem if you are eager to do so |
| 2 | You have been a good reader all the time |
| 3 | It is impossible to increase one's problem solving ability since we have limited IQs |
| 4 | You were more close to your Mom than your Dad |
| 5 | To be creative, one must apply consistent effort to problems encountered |
| 6 | In comparison to your friends, you normally day dream more |
| 7 | It is always best to be under some stress when solving a problem |
| 8 | One is likely to be more creative and imaginative, if he/she possesses more intelligence |
| 9 | Building confidence through repeated success will always increase one's problem solving ability |
| 10 | It is best to strongly focus one's total attention on the problem and try to think it through |

provided in Ref. 8. Interpretations of correct answers to these ten items are as follows[8]:

- **Correct answers to items 7–10:** You are a creative individual and should be working in a job environment that clearly offers you the opportunity to be an "idea person".
- **Correct answers to items 3–6:** You are one of the vast majority of individuals who possess average ability to generate imaginative solutions to problems.
- **Correct answer to items 0–2:** You are not an imaginative individual and should work in areas that require you to follow closely a set plan of actions.

#### 11.2.2.2 *Test II*

This is another test that is designed to assess the degree of creativity a person possesses. The test is composed of fifty statements.[9] The individual taking the test is required to provide one of the five possible answers (i.e., $A$ = strongly agree, $B$ = agree, $C$ = in between or don't know, $D$ = disagree, or $E$ = strongly disagree) to each of these fifty statements. Answers $A$ and $E$ are given either 2 or $-2$ marks. Similarly, answers $B$ and $D$ are given either 1 or $-1$ mark. Finally, a zero mark is given to answers $C$. A sample of statements covered in the test is presented in Table 11.2.[9]

Numerical scores for each of the fifty test statements are given in Ref. 9. The total score for each individual is obtained by adding up the score given

Table 11.2. A sample of the test statements.

| No. | Statement |
|---|---|
| 1 | I consider that it would be a waste of time to ask questions, if there is no hope of obtaining answers |
| 2 | Sometime, I get rather overly enthusiastic about things |
| 3 | Day dreaming has resulted in the impetus for many of my important projects |
| 4 | I possess a quite high degree of aesthetic sensitivity |
| 5 | I like the type of work in which I influence other people |
| 6 | In my opinion, people who strive for perfection are unwise |
| 7 | I feel that in evaluating information, the source of information is more important than its contents |
| 8 | I avoid asking questions that show ignorance |
| 9 | In my opinion, hard work is the basic factor in success |
| 10 | Inability to find solution to a problem is often due to asking incorrect questions |
| 11 | It is totally waste of time to perform analysis of one's failures |
| 12 | The main problem with many individuals is that they take things to heart |

to each test statement. The total score for each individual with respect to his/her creativeness is interpreted as follows[9]:

- −100 to 19: Noncreative
- 20 to 30: Below average
- 40 to 59: Average
- 60 to 79: Above average
- 80 to 100: Very creative.

## 11.3 Creativity Recording

Recording of inventions and ideas is as important as thinking them at the first place. Records are important for various purposes. For example, with respect to inventions, records are necessary for the purposes such as follows[3]:

- To antedate an earlier filed patent or a publication that describes partially the invention one is seeking to protect.
- To avoid liability for infringing.
- To determine the right inventorship.
- To antedate the invention date of a competing applicant for a patent on the invention in question.
- To establish one's prior knowledge concerning the invention which another party claims to have pass on (i.e., to the individual) in confidence.
- To establish prior knowledge of an invention, before the hiring date of an employee who may be bringing in a comparable idea for an invention with him/her.

### 11.3.1 *Tools for Recording Ideas*

There are many tools that can be used to record ideas. Ten of these tools are shown in Fig. 11.2.[4] Each of the tools is described in detail, along with its advantages and disadvantages, in Ref. 4.

## 11.4 Patents

The term "Patent" may mean different things to different people. For example, the issued patent document, a contract between the government and the inventor, the act of applying for a patent, or a time limited monopoly granted by the government to the inventor.[5]

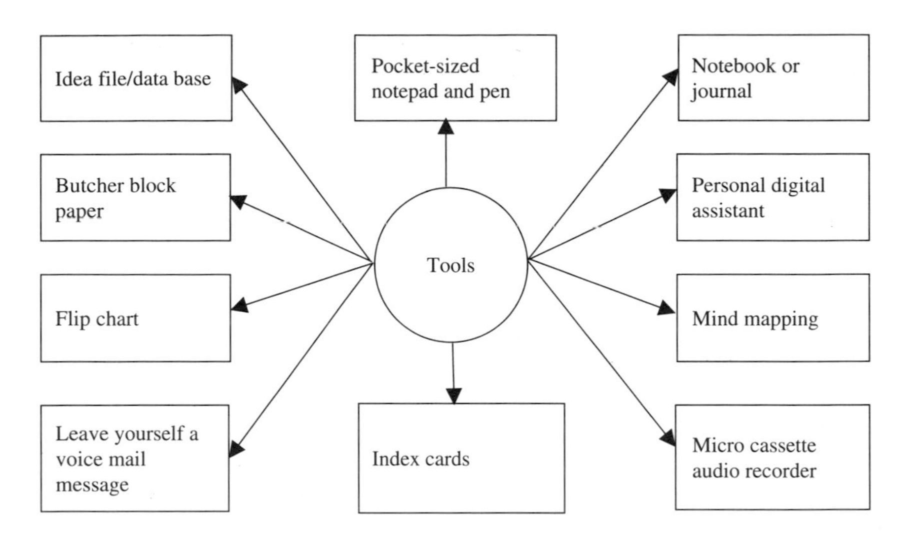


Fig. 11.2. Tools for recording ideas.

The history of Patents may be traced back to the passing of the British Statute of Monopolies by the parliament in 1623. This act was the prototype of modern patent statutes in the English speaking countries. In early America, as there were no general laws for the granting of patents, the inventors were required to appeal to the governing body of their Colony or State. In 1641, the Massachusetts General Court granted the first such patent to Samuel Winslow for a novel method of making salt.[5]

### 11.4.1 *What Can be Patented or Cannot be Patented?*

In the United States, a patent may be granted for a new and useful machine (i.e., any apparatus with moving parts. An electronic apparatus is also included under this category), manufacture (i.e., article without moving parts), process (i.e., primarily industrial or technical), plant (i.e., the one has been asexually propagated), or composition of matter (i.e., the one relates to chemical compositions).[5] These patents are often referred to as "utility patents".

Patent laws in the United States also have a provision for granting "design patents". A design patent can be granted to anyone who has invented an ornamental, new, and original design for an item of manufacture. This type of patent protects only the appearance of an item and is granted for 3, 5, 7, or 14 years.[5]

All inventions are not patentable. Products of nature, mental processes, systems of logic, and methods of doing business cannot be patented. Also, inventions useful specifically in the utilization of atomic energy for atomic weapons or special nuclear material cannot be patented.

### 11.4.2 *Who May Apply for a Patent? Is Filing Time Important?*

Usually, only the true inventor(s) can apply for a patent. True inventors are the individuals who furnish ideas of the patent (i.e., not the employer or the people who make the money investment). Patent laws do not discriminate age or sex. Furthermore, a foreign national can also obtain a US patent under exactly the same terms and conditions applicable to American citizens.

A corporate engineer or others whose invention is "assigned" to the company, in the legal sense file the application for the patent. However, the "rights" to the invention still belong to the company itself.[5]

Patent filing time is important as the Patent Office, assuming no other complications, grants a patent to the first inventor. It is impossible to find out whether a competing inventor come up with the invention first or filed his/her application first. It means that there is a definite need for urgency in filing for a patent. Nonetheless, the latest, one can file a patent application is one year after making the invention public in a printed form.

### 11.4.3 *Patent Attorney Selection and Functions*

This professional (i.e., Patent Attorney) acts as a communicator between the inventor and the Patent Office.[5] In the selection of a Patent Attorney, a careful consideration must be given to the possession of two diverse skills by the selected individual.[3] The first is the expertise in writing patent applications and licensed by the US Patent Office to perform this task. The second is the expertise in the invention's area of technology.

The Patent Attorney performs various functions including filing the patent application, making patent searches, preparing the patent application, helping to decide whether to file the patent in US only, foreign countries, or both, and following-up on the patent application.[5]

### 11.4.4 *The Patenting Process*

The patenting process is composed of many steps as shown in Fig. 11.3.[10]

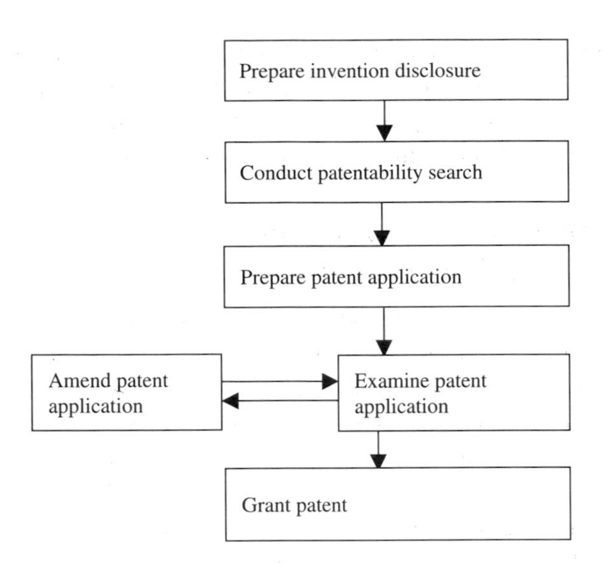


Fig. 11.3. The patenting process steps.

The step "Prepare invention disclosure" is concerned with preparing a disclosure of the invention in written form. The disclosure provides various types of information concerning the invention including what the invention is, how it works, how it can be utilized, and how it is an improvement over already existing articles/methods/approaches. Also, disclosure usually includes some drawings of the invention.

The step "Conduct patentability search" is concerned with carrying out a patentability search in the Untied States Patent and Trademark Office in Washington, D.C. Its main purpose is to decide whether to proceed with the patent application or not. Nonetheless, through this search, all properly classified and filed US patents pertinent to the invention are reviewed and the past experiences indicate that normally patentability search is around 90% effective in locating any pertinent related material to the invention.

The step "Prepare patent application" is concerned with preparing the application for obtaining the patent. The preparation of application depends on various factors including the technical complexity of the subject matter, the quality of inventor's written description of the invention, and the number of application revisions necessitated by defining the invention by the inventor during the application drafting process.

The step "Examine patent application" is concerned with examining of the patent application by the United States Patent and Trademark

Office. The office appoints a patent examiner for determining the invention patentability with respect to usefulness, novelty, and nonobviousness. The typical time span between patent application filing and receipt of the first Office Action is 12–24 months.

The step "Amend patent application" is concerned with amending the patent application as per the United States Patent and Trademark Office "Action" set forth, after the application was reviewed, with respect to patentability.

The step "Grant patent" is concerned with granting the patent when the results of the patent examination are favorable. After the successful issuance of the patent, maintenance fees are paid at 3.5, 7.5, and 11.5 years from the date of patent issue for keeping the patent in force.

## Problems

1. Define a patent.
2. Write an essay on creativity testing.
3. Discuss the following terms with respect to creative tests:
   - Divergent thinking
   - Convergent thinking
4. Write down a list of statements useful for testing the creative ability of a person.
5. Discuss the following two creativity tests:
   - Wallas and Kogan Test
   - Guilford's alternative uses task test.
6. Discuss the need for creativity recording.
7. List at least ten useful tools for recording ideas.
8. Discuss the functions of a Patent Attorney.
9. Discuss the patenting process steps.
10. Discuss "What can be patented or cannot be patented".

## References

1. Raudsepp, E., Testing for Creativity, *Machine Design*, Vol. 37, No. 15, 24 June 1965, pp. 122–128.
2. How creative are you?, Princeton Creative Research, Inc., 10 Nassau Street, P.O. Box 122, Princeton, New Jersey, USA.

3. Middendorf, W.H., *What Every Engineer Should Know About Inventing?*, Marcel Dekker, Inc., New York, 1981.
4. Frey, C., Ten Power Tools for Recording Your Ideas, Innovation Tools, December 2003. Available online at http://www.innovationtools.com/Articles/ArticleDetails.asp?a=113.
5. Bailey, R.L., *Disciplined Creativity for Engineers*, Ann Arbor Science Publishers, Inc., Ann Arbor, Michigan, 1978.
6. Gilchrist, M., *The Psychology of Creativity*, Melbourne University Press, Melbourne, Australia, 1972.
7. Guilford, J.P., Creativity, *American Psychologist*, Vol. 5, 1950, pp. 444–454.
8. Gillis, J.G., Creativity, Problem Solving and Decision Making, in *Creativity*, edited by Dale Timpe, A., Facts on File Publications, New York, 1987, pp. 209–213.
9. Raudsepp, E., How Creative are You?, in *Creativity*, edited by Dale Timpe, A., Facts on File Publications, New York, 1987, pp. 34–38.
10. Colitz, M.J., Overview of the Patenting Process, The Law Office of Edward P. Dutkiewicz, 640 Douglas Avenue, Dunedin, Florida. Also, available online at http://www.colitz.com/site/.

## Appendix

# Bibliography: Literature on Creativity

## A.1 Introduction

Over the years, many publications directly or indirectly related to creativity in engineering have appeared in the form of journal articles, books, conference proceedings articles, technical reports, and so on. This Appendix presents an extensive list of such publications.

The period covered by the listing is from 1958–2004. The main objective of this listing is to provide readers with sources for obtaining additional information on creativity directly or indirectly related to engineering.

## A.2 Publications


1. Abdelhay, S.A., Creativity in Computer Control, *Intech*, Vol. 34, No. 2, 1987, pp. 49–51.
2. Adams, R. (editor), *Creativity in Communications*, Studio Vista, London, 1971.
3. Addis, W., *Creativity and Innovation: The Structural Engineer's Contribution to Design*, Architectural Press, Boston, 2001.
4. Ahmed, A.M. and Abdalla, H.S., Role of Innovation Process in Crafting the Vision of the Future, *Computers and Industrial Engineering*, Vol. 37, No. 1, 1999, pp. 421–424.
5. Aihara, K. and Hori, K., Enhancing Creativity Through Reorganizing Mental Space Concealed in a Research Notes Stack, *Knowledge-Based Systems*, Vol. 11, Nos. 7–8, 1998, pp. 469–478.

6. Akin, O. and Akin, C., On the Process of Creativity in Puzzles, Inventions, and Designs, *Automation in Construction*, Vol. 7, Nos. 2–3, 1998, pp. 123–138.
7. Amabile, T.M. and Gryskiewicz, S.S., *Creativity in the R & D Laboratory*, Center for Creative Leadership, Greensboro, North Carolina, 1991.
8. Amabile, T.M., How Work Environments Affect Creativity, *Proceedings of the IEEE International Conference on Systems, Man and Cybernetics*, 1997, pp. 50–55.
9. Amabile, T., *Creativity in Context*, Westview Press, Boulder, Colorado, 1996.
10. Amabile, T.M. and Conti, R., Changes in the Work Environment for Creativity During Downsizing, *Academy of Management Journal*, Vol. 42, No. 6, 1999, pp. 630–640.
11. Amabile, T.M., Hadley, C.N. and Kramer, S.J., Creativity Under the Gun, *Harvard Business Review*, Vol. 80, No. 8, 2002, pp. 52–60.
12. Amin, M., Development and Leadership of Research Consortia: Lessons Learned and Possible Road Ahead for Continued Innovation, *Proceedings of the IEEE Power Engineering Society Transmission and Distribution Conference*, Vol. 3, 2002, pp. 1710–1715.
13. Amin, M., Development and Leadership of Research Consortia: Lessons Learned and Possible Road Ahead for Continued Innovation, *Proceedings of the IEEE Power Engineering Society Transmission and Distribution Conference*, Vol. 2, 2001, pp. 1085–1091.
14. Amoroso, D.L. and Eriksson, I.V., Use of Content Analysis for Studying the Creativity Construct in the Context of Technology-Rich Applications, *Proceedings of the Hawaii International Conference on System Sciences*, 2000, pp. 193–195.
15. Andriopoulos, C., Six Paradoxes in Managing Creativity: An Embracing Act, *Long Range Planning*, Vol. 36, No. 4, 2003, pp. 375–388.
16. Amidon, D.M., The Creativity Challenge: Management of Innovation and Technology, *Research-Technology Management*, Vol. 39, No. 3, 1996, pp. 60–65.
17. Anonk, G., Managing and Manipulating Creative Ideas, *Folding Carton Industry*, Vol. 26, No. 6, 1999, pp. 34–35.
18. Arasteh, A.R., *Creativity in the Life-Cycle*, E.J. Brill, Leiden, The Netherland, 1968.
19. Arieti, S., *Creativity: The Magic Synthesis*, Basic Books, New York, 1976.
20. Armbrecht, F.M.R. *et al.*, Knowledge Management in Research and Development, *Research-Technology Management*, Vol. 44, No. 4, 2001, pp. 28–48.
21. Azani, H. and Khorramshahgol, R., Impact of Automation on Engineers' Creativity and Innovation and Its Implications for Reducing Resistance to Change, *Computers in Industry*, Vol. 16, No. 4, 1991, pp. 377–383.
22. Baer, J., *Creativity and Divergent Thinking: A Task-Specific Approach*, L. Erlbaum Associates, Hillsdale, New Jersey, 1993.
23. Bailey, R.L., *Disciplined Creativity for Engineers*, Ann Arbor Science Publishers, Ann Arbor, Michigan, 1978.

24. Baillie, C., Enhancing Creativity in Engineering Students, *Engineering Science and Education Journal*, Vol. 11, No. 5, 2002, pp. 185–192.
25. Bangle, C., The Ultimate Creativity Machine — How BMW Turns Art into Profit, *Harvard Business Review*, Vol. 79, No. 1, 2001, pp. 47–53.
26. Barak, M. and Goffer, N., Fostering Systematic Innovative Thinking and Problem Solving: Lessons Education can Learn from Industry, *International Journal of Technology and Design Education*, Vol. 12, No. 3, 2002, pp. 227–247.
27. Barlow, C., Insight or Ideas: Escaping the Idea Centered "Box" Defining Creativity, *Proceedings of the Hawaii International Conference on System Sciences*, 2001, p. 209.
28. Barron, F.X., *Creativity and Personal Freedom*, Van Nostrand, Princeton, New Jersey, 1968.
29. Basadur, M. and Hausdorf, P.A., Measuring Divergent Thinking Attitudes Related to Creative Problem Solving and Innovation Management, *Creativity Research Journal*, Vol. 9, No. 1, 1996, pp. 21–32.
30. Bawden, D., Information Systems and the Stimulation of Creativity, *Journal of Information Science*, Vol. 12, No. 5, 1986, pp. 203–216.
31. Becker, J.D., Creativity and Precision in Evolutionary Synthesis and the Hierarchical Selection Scheme, *Chaos Solitons and Fractals*, Vol. 11, No. 1, 2000, pp. 345–350.
32. Benami, O. and Jin, Y., Creative Stimulation in Conceptual Design, *Proceedings of the ASME Design Engineering Technical Conference*, Vol. 4, 2002, pp. 251–263.
33. Benton, M.C., Creativity in Research and Invention in the Physical Sciences: An Annotated Bibliobraphy, U.S. Naval Research Laboratory, Washington, D.C., 1961.
34. Bharadwaj, S. and Menon, A., Making Innovation Happen in Organizations: Individual Creativity Mechanisms, Organizational Creativity Mechanisms or Both?, *Journal of Product Innovation Management*, Vol. 17, No. 6, 2000, pp. 424–434.
35. Bichard, M., Creativity, Leadership and Change, *Public Money and Management*, Vol. 20, No. 2, 2000, pp. 41–46.
36. Blosiu, J.O., Use of Synectics as an Idea Seeding Technique to Enhance Design Creativity, *Proceedings of the IEEE International Conference on Systems, Man and Cybernetics*, Vol. 3, 1999, pp. III-1001–III-1006.
37. Boden, M.A., Agents and Creativity, *Communications of the ACM*, Vol. 37, No. 7, 1994, pp. 117–121.
38. Boden, M.A., Creativity and Artificial Intelligence, *Artificial Intelligence*, Vol. 103, Nos. 1–2, 1998, pp. 347–356.
39. Boden, M.A., Creativity and Computers, *Cybernetics and Systems*, Vol. 26, No. 3, 1995, pp. 267–293.
40. Bond, P. and Otterson, P., Creativity Enhancement Software: A Systemic Approach, *International Journal of Technology Management*, Vol. 15, Nos. 1–2, 1998, pp. 173–191.

41. Bonnardel, N., Towards Understanding and Supporting Creativity in Design: Analogies in a Constrained Cognitive Environment, *Knowledge-Based Systems*, Vol. 13, Nos. 7–8, 2000, pp. 505–513.
42. Bonner, J.M., Ruekert, R.W. and Walker, O.C., Upper Management Control of New Product Development Projects and Project Performance, *Journal of Product Innovation Management*, Vol. 19, No. 3, 2002, pp. 233–245.
43. Borenstein, A., *Creativity in Later Life: A Selected, Annotated Bibliography*, Garland, New York, 1985.
44. Bouchard, C. and Aoussat, A., Design Process Perceived as an Information Process to Enhance the Introduction of New Tools, *International Journal of Vehicle Design*, Vol. 31, No. 2, 2003, pp. 162–175.
45. Bringsjord, S., Bello, P. and Ferrucci, D., Creativity, the Turing Test, and the (Better) Lovelace Test, *Minds and Machines*, Vol. 11, No. 1, 2001, pp. 3–27.
46. Brockman, B. and Morgan, R.M., The Role of Existing Knowledge in New Product Innovativeness and Performance, *Decision Sciences*, Vol. 34, No. 2, 2003, pp. 385–419.
47. Brockman, J. (editor), *Creativity*, Simon & Schuster, New York, 1993.
48. Buchanan, B.G., Creativity at the Metalevel AAAI-2000 Presidential Address, *AI Magazine*, Vol. 22, No. 3, 2001, pp. 13–28.
49. Buchsbaum, S.J., Managing Creativity: For Fun and for Profit, *International Journal of Technology Management*, Vol. 1, Nos. 1–2, 1986, pp. 51–64.
50. Caldwell, D.F. and O'Reilly, C.A., The Determinants of Team-Based Innovation in Organizations — The Role of Social Influence, *Small Group Research*, Vol. 34, No. 4, 2003, pp. 497–517.
51. Candy, L. and Edmonds, E. (editors), *Proceedings of the Third Creativity and Cognition Conference*, ACM Press, New York, 1999.
52. Carland, J.C., *Creativity and Innovation in Business: Requisites for Success*, Dame Pub., Houston, 1997.
53. Carlson, L.E. and Sullivan, J.F., Engineers Invent and Innovate, *Proceedings of the Frontiers in Education Conference*, Vol. 2, 2000, pp. F2G-11–F2G-15.
54. Carter, D.E., Role of Test in Design Creativity, *Computer-Aided Engineering Journal*, Vol. 4, No. 3, 1987, pp. 137–139.
55. Cassidy, J.E., From "Half-Brain" to "Whole Brain": Learn to Create High Performing Teams, *Proceedings of the Annual Quality Congress*, 1998, pp. 725–735.
56. Chen, Z., Combining Creativity and Expertise, *Cybernetics and Systems*, Vol. 28, No. 4, 1997, pp. 327–336.
57. Clark, K. and James, K., Justice and Positive and Negative Creativity, *Creativity Research Journal*, Vol. 12, No. 4, 1999, pp. 311–320.
58. Clegg, B., *Creativity and Innovation for Managers*, Butterworth-Heinemann, Boston, 1999.
59. Coates, J.F., Innovation in the Future of Engineering Design, *Technological Forecasting and Social Change*, Vol. 64, Nos. 2–3, 2000, pp. 121–132.
60. Colonna, J.F., Creativity and Simplicity, *Visual Computer*, Vol. 11, No. 8, 1995, pp. 447–452.

61. Colwill, J., Innovation Plus Creativity is Key, *Packaging Week*, Vol. 6, No. 9, 2003, p. 23.
62. Computer-Aided Task Force, Creativity and Architecture: The Impact of New Electronic Tools, American Institute of Architects, Washington, DC, 1993.
63. Conceicao, P. and Heitor, M.V., Knowledge Interaction Towards Inclusive Learning: Promoting Systems of Innovation and Competence Building, *Technological Forecasting and Social Change*, Vol. 69, No. 7, 2002, pp. 641–651.
64. Cooper, R.B., Information Technology Development Creativity: A Case Study of Attempted Radical Change, *MIS Quarterly*, Vol. 24, No. 2, 2000, pp. 245–276.
65. Couger, J.D., *Creativity and Innovation in Information Systems Organizations*, Boyd & Fraser, Hinsdale, Illinois, 1996.
66. Couger, J.D. and Dengate, G., Measurement of Creativity of I.S. Products, *Information Systems Proceedings of the Hawaii International Conference on System Science*, Vol. 4, 1992, pp. 288–298.
67. Couger, J.D., Measurement of the Climate for Creativity in I.S. Organizations, *Proceedings of the Hawaii International Conference on System Sciences*, Vol. 4, 1994, pp. 351–357.
68. Creativity and Conformity, a Problem for Organizations, Foundation for Research on Human Behavior, Ann Arbor, Michigan, 1958.
69. Cropley, A.J., *Creativity*, Longmans, London, 1968.
70. Cropley, A.J. and Dehn, D. (editors), *Fostering the Growth of High Ability: European Perspectives*, Ablex Publ. Corp., Norwood, New Jersey, 1996.
71. Cropley, D.H. and Cropley, A.J., Fostering Creativity in Engineering Undergraduates, *High Ability Studies*, Vol. 11, No. 2, 2000, pp. 207–219.
72. Crosby, A.C., *Creativity and Performance in Industrial Organization*, Tavistock Publications, New York, 1968.
73. Cunha, M.P.E., da Cunha, J.V. and Dahab, S., Yin-Yang: A Dialectical Approach to Total Quality Management, *Total Quality Management*, Vol. 13, No. 6, 2002, pp. 843–853.
74. Dahlgaard, J.J. and Dahlgaard, S.M.P., Integrating Business Excellence and Innovation Management: Developing a Culture for Innovation, Creativity and Learning, *Total Quality Management*, Vol. 10, Nos. 4–5, 1999, pp. S465–S472.
75. Dale Timpe, A. (editor), *Creativity*, Facts on File Publications, New York, 1987.
76. Dartnall, T. (editor), *Creativity, Cognition, and Knowledge: An Interaction*, Praeger, Westport, Connecticut, 2002.
77. Dasgupta, S., *Creativity in Invention and Design: Computational and Cognitive Explorations of Technological Originality*, Cambridge University Press, New York, 1994.
78. Dauw, D.C. and Fredian, A.J., *Creativity and Innovation in Organizations*, Kendall/Hunt Pub. Co., Dubuque, Iowa, 1976.

79. Davies, L.J. and Ledington, P.W.J., Creativity and Metaphor in Soft Systems Methodology, *Journal of Applied Systems Analysis*, Vol. 15, 1988, pp. 31–36.
80. Davis, G.A., *Creativity is Forever*, Kendall/Hunt Pub., Dubuque, Iowa, 1998.
81. De Dreu, C.K.W. and West, M.A., Minority Dissent and Team Innovation: The Importance of Participation in Decision Making, *Journal of Applied Psychology*, Vol. 86, No. 6, 2001, pp. 1191–1201.
82. DeGraff, J.T. and Lawrence, K.A., *Creativity at Work: Developing the Right Practices to Make Innovation Happen*, Wiley, New York, 2002.
83. Deutsch and Shea, Inc., Creativity: A Comprehensive Bibliography on Creativity in Science, Engineering, Business, and the Arts, Industrial Relations News, New York, 1958.
84. DeVore, H.L., *Creativity, Design and Technology*, Davis Publications, Worcester, Massachusetts, 1989.
85. Dewett, T., Understanding the Relationship Between Information Technology and Creativity in Organizations, *Creativity Research Journal*, Vol. 15, Nos. 2–3, 2003, pp. 167–182.
86. Di Cyan, E., *Creativity: Road to Self-Discovery*, Jove Publications, New York, 1978.
87. Dickey, J.W., *Cyberquest: Problem Solving and Innovation Support System Conceptual Background and Experiences*, Ablex Publ. Corp., Norwood, New Jersey, 1995.
88. Dolby, R.E., Management Processes for Innovation in Joining Technology, *Welding Research Abroad*, Vol. 48, No. 12, 2002, pp. 23–25.
89. Drabkin, S., Enhancing Creativity When Solving Contradictory Technical Problems, *Journal of Professional Issues in Engineering Education and Practice*, Vol. 122, No. 2, 1996, pp. 78–82.
90. Edmonds, E., Fischer, G., Mountford, S.J., Nake, F., Riecken, D. and Spence, R., Creativity: Interacting with Computers, *Proceedings of the Conference on Human Factors in Computing Systems*, Vol. 2, 1995, pp. 185–186.
91. Ekvall, G., *Creativity at the Place of Work. A Study of Suggestors and Suggestion Systems in the Swedish Mechanical Industry*, PA-radet, Stockholm, 1971.
92. Elwes, A. (editor), *Creativity Works*, Profile Books, London, 2000.
93. Ensor, J., Cottam, A. and Band, C., Fostering Knowledge Management Through the Creative Work Environment: A Portable Model from the Advertising Industry, *Journal of Information Science*, Vol. 27, No. 3, 2001, pp. 147–155.
94. Evans, A., Housing Management — A Guide to Quality and Creativity — Power, A, *Environment and Planning B-Planning and Design*, Vol. 19, No. 4, 1992, pp. 487–488.
95. Evans, J.R., Creativity in MS/OR — Overcoming Barriers to Creativity, *Interfaces*, Vol. 23, No. 6, 1993, pp. 101–106.

96. Farid, F., El-Sharkawy, A.R. and Austin, L.K., Managing for Creativity and Innovation in A/E/C Organizations, *Journal of Management in Engineering*, Vol. 9, No. 4, 1993, pp. 399–409.
97. Farris, D.F., Imagination Engineering: A Toolkit for Business Creativity, *Library Journal*, Vol. 121, No. 16, 1996, p. 94.
98. Fellers, G., *Creativity for Leaders*, Pelican Pub. Co., Gretna, Louisiana, 1996.
99. Feurer, R., Chaharbaghi, K. and Wargin, J., Developing Creative Teams for Operational Excellence, *International Journal of Operations and Production Management*, Vol. 16, No. 1, 1996, p. 5.
100. Filion, L.J., Visionary Systems Thinking (VST) as a Support to Creativity in the Quality Management (TQM) Process, *Systems Research*, Vol. 11, No. 1, 1994, pp. 125–133.
101. Fischer, G., Turning Breakdowns into Opportunities for Creativity, *Knowledge-Based Systems*, Vol. 7, No. 4, 1994, pp. 221–232.
102. Fischer, S., Boogaard, M. and Huysman, M., Innovation and Creativity versus Control Assessing the Paradox in IT Organizations, *Proceedings of the Hawaii International Conference on System Sciences*, Vol. 4, 1994, pp. 367–376.
103. Fodor, E.M. and Carver, R.A., Achievement and Power Motives, Performance Feedback, and Creativity, *Journal of Research in Personality*, Vol. 34, No. 4, 2000, pp. 380–396.
104. Fodor, E.M. and Roffe-Steinrotter, D., Rogerian Leadership Style and Creativity, *Journal of Research in Personality*, Vol. 32, No. 2, 1998, pp. 236–242.
105. Ford, C.M. and Gioia, D.A., Factors Influencing Creativity in the Domain of Managerial Decision Making, *Journal of Management*, Vol. 26, No. 4, 2000, pp. 705–732.
106. Fraser, R.A. and Voegtlen, D., Creativity. Method or Magic?, *SAVE Proceedings* (Society of American Value Engineers), Vol. 25, 1996, pp. 3–16.
107. Freeman, J., *Creativity: A Selective Review of Research*, Society for Research into Higher Education, London, 1971.
108. Galletta, D.F., Sampler, J.L. and Teng, J.T.C., Strategies for Integrating Creativity Principles into the System Development Process, *Proceedings of the Hawaii International Conference on System Science*, Vol. 4, 1992, pp. 268–278.
109. Gallivan, M.J., The Influence of Software Developers' Creative Style on Their Attitudes to and Assimilation of a Software Process Innovation, *Information Management*, Vol. 40, No. 5, 2003, pp. 443–465.
110. Gamache, R.D. and Kuhn, R.L., *The Creativity Infusion: How Managers can Start and Sustain Creativity and Innovation*, Harper & Row, New York, 1989.
111. Gamez, G., *Creativity: How to Catch Lightning in a Bottle*, Peak Publications, Los Angeles, 1996.
112. Gedo, J.E. and Gedo, M.M., *Perspectives on Creativity: The Biographical Method*, Ablex Publ. Corp., Norwood, New Jersey, 1992.

113. Gero, J.S., Creativity, Emergence and Evolution in Design, *Knowledge-Based Systems*, Vol. 9, No. 7, 1996, pp. 435–448.
114. Ginn, M.E., *The Creativity Challenge: Management of Innovation and Technology*, JAI Press, Greenwich, Connecticut, 1995.
115. Ginn, M.E., Creativity Management — Systems and Contingencies from a Literature-Review, *IEEE Transactions on Engineering Management*, Vol. 33, No. 2, 1986, pp. 96–101.
116. Glass, R.L., A Story About the Creativity Involved in Software Work, *IEEE Software*, Vol. 18, No. 5, 2001, pp. 96–97.
117. Glassman, E., *The Creativity Factor: Unlocking the Potential of your Team*, Pfeiffer & Co., San Diego, 1991.
118. Glynn, M.A., Innovative Genius: A Framework for Relating Individual and Organizational Intelligences to Innovation, *Academy of Management Review*, Vol. 21, No. 4, 1996, pp. 1081–1111.
119. Goel, P.S. and Singh, N., Creativity and Innovation in Durable Product Development, *Computers and Industrial Engineering*, Vol. 35, Nos. 1–2, 1998, pp. 5–8.
120. Goldenberg, J. and Mazursky, D., *Creativity in Product Innovation*, Cambridge University Press, New York, 2002.
121. Goman, C.K., *Creativity in Business: A Practical Guide for Creative Thinking*, Crisp Publications, Menlo Park, California, 2000.
122. Gowan, J.C., *Creativity: Its Educational Implications*, Wiley, New York, 1967.
123. Gowan, J.C., Khatena, J. and Torrance, E.P., *Creativity: Its Educational Implications*, 2nd edn., Kendall/Hunt Pub. Co., Dubuque, Iowa, 1981.
124. Grant, J.M., *Creativity in Motion*, Learning Publications, Holmes Beach, Florida, 1992.
125. Gupta, A.K. and Singhal, A., Managing Human-Resources for Innovation and Creativity, *Research-Technology Management*, Vol. 36, No. 3, 1993, pp. 41–48.
126. Gutierrez, O. and Greenberg, E., Creative Problem-Solving in the Specification of Information Requirements, *Systems Practice*, Vol. 6, No. 6, 1993, pp. 647–667.
127. Haapasalo, H. and Kess, P., Managing Creativity: Is it Possible to Control the Birth of Innovation in Product Design?, *International Journal of Technology Management*, Vol. 24, No. 1, 2002, pp. 57–69.
128. Haefele, J.W., *Creativity and Innovation*, Reinhold, New York, 1962.
129. Hafter, D.M., The Cost of Inventiveness — Labor's Struggle with Management's Machine, *Technology and Culture*, Vol. 44, No. 1, 2003, pp. 102–113.
130. Hall, D.J., Role of Creativity Within Best Practice Manufacturing, *Technovation*, Vol. 16, No. 3, 1996, pp. 115–121.
131. Hare, A.P., *Creativity in Small Groups*, Sage Publications, Beverly Hills, California, 1982.
132. Harman, W.W., *Creativity and Intuition in Business: The Unconscious Mind and Management Effectiveness*, SRI International, Menlo Park, California, 1985.

133. Harrington, H.J., Hoffherr, G.D. and Reid, R.P., *The Creativity Toolkit: Provoking Creativity in Individuals and Organizations*, McGraw-Hill, New York, 1997.
134. Harvey, M., Novicevic, M.M. and Kiessling, T., Development of Multiple IQ Maps for Use in the Selection of Inpatriate Managers: A Practical Theory, *International Journal of Intercultural Relations*, Vol. 26, No. 5, 2002, pp. 493–524.
135. Heck, P.S. and Ghosh, S., Study of Synthetic Creativity: Behavior Modeling and Simulation of an Ant Colony, *IEEE Intelligent Systems and Their Applications*, Vol. 15, No. 6, 2000, pp. 58–66.
136. Heinzen, T.E., *Everyday Frustration and Creativity in Government: A Personnel Challenge to Public Administration*, Ablex Publ. Corp., Norwood, New Jersey, 1994.
137. Helle, P.F., Creativity: The Key to Breakthrough Changes, How Teaming can Harness Collective Knowledge, *Proceedings of the Annual American Production and Inventory Control Society International Conference*, 1997, pp. 301–303.
138. Hender, J., Dean, D., Rodgers, T. and Nunamaker, J, Improving Group Creativity: Brainstorming versus Non-Brainstorming Techniques in a GSS Environment, *Proceedings of the Hawaii International Conference on System Sciences*, 2001, p. 37.
139. Hendrix, V.E., Chapla, D.B. and Mizzelle, W. (editors), *Proceedings of the Creativity and Innovation Symposium*, National Defense University, Washington, D.C., 1985.
140. Henry, J., *Creativity and Perception in Management*, SAGE Publications, Thousand Oaks, California, 2001.
141. Hewett, T.T. and Kavanagh, T. (editors), *Proceedings of the Fourth Creativity and Cognition Conference*, ACM Press, New York, 2002.
142. Hiam, A., *The Manager's Pocket Guide to Creativity*, HRD Press, Amherst, Massachusetts, 1998.
143. Hicks, B.J. *et al.*, A Framework for the Requirements of Capturing, Storing and Reusing Information and Knowledge in Engineering Design, *International Journal of Information Management*, Vol. 22, No. 4, 2002, pp. 263–280.
144. Higgins, L.F., Applying Principles of Creativity Management to Marketing Research Efforts in High-Technology Markets, *Industrial Marketing Management*, Vol. 28, No. 3, 1999, pp. 305–317.
145. Higgins, L.F., Couger, J.D. and McIntyre, S.C., Creative Approaches to Development of Marketing Information Systems, *Proceedings of the Hawaii International Conference on System Sciences*, 1992, pp. 398–404.
146. Hinds, P.J., Hidden Cost of Keeping Secrets: How Protecting Proprietary Information can Inhibit Creativity, *Proceedings of the Hawaii International Conference on System Sciences*, 2000, p. 172.
147. Hinrichs, J.R., *Creativity in Industrial Scientific Research: A Critical Survey of Current Opinion, Theory, and Knowledge*, American Management Association, New York, 1961.

148. Hirota, R. and Ishikawa, A., ISOP: A Computerized Group-Based Intelligent Problem Solving Technique, *Proceedings of the Hawaii International Conference on System Sciences*, Vol. 4, 1994, pp. 291–299.
149. Hodges, J., Flowers, M. and Dyer, M., Knowledge Representation for Design Creativity, *American Society of Mechanical Engineers, Production Engineering Division (Publication) PED*, Vol. 25, 1996, pp. 81–93.
150. Horowitz, R.S., Latinwo, L. and Weiner, J.M., Building Your Own Database: Potential for Creativity, *Proceedings of the ASIS Annual Meeting*, Vol. 24, 1987, pp. 106–110.
151. Hunter, J., Industry Told it Must Focus on Creativity, *Packaging Week*, Vol. 6, No. 1, 2003, p. 3.
152. Ingalls, R., *Creativity for Life: My Story and My Work*, Log Cabin Press, Woodstock, New York, 1982.
153. Isaksen, S.G. *et al.* (editors), *The Emergence of a Discipline*, Ablex Publ. Corp., Norwood, New Jersey, 1993.
154. Ishihama, M., Training Students on the TRIZ Method Using a Patent Database, *International Journal of Technology Management*, Vol. 25, Nos. 6–7, 2003, pp. 568–578.
155. Isova, V., Creativity in the Management — Gulas, S, *Ekonomicky Casopis*, Vol. 35, No. 1, 1987, pp. 88–89.
156. Itoh, T., Abduction for Creativity, *International Journal of Technology Management*, Vol. 25, Nos. 6–7, 2003, pp. 507–516.
157. Ivanicka, K., Creativity as an Economic Category and its Management in the Light of Synergetics, *Ekonomicky Casopis*, Vol. 36, No. 9, 1988, pp. 795–810.
158. Jessop, J.L.P., Expanding our Students' Brainpower: Idea Generation and Critical Thinking Skills, *IEEE Antennas and Propagation Magazine*, Vol. 44, No. 6, 2002, pp. 140–144.
159. Joseph, D., Ang, S. and Slaughter, S., Soft Skills and Creativity in IS Professionals, *Proceedings of the Hawaii International Conference on System Sciences*, 1999, pp. 234–235.
160. Judge, W.Q., Fryxell, G.E. and Dooley, R.S., The New Task of R & D Management: Creating Goal-Directed Communities for Innovation, *California Management Review*, Vol. 39, No. 3, 1997, p. 72.
161. Kappel, T.A. and Rubenstein, A.H., Creativity in Design: The Contribution of Information Technology, *IEEE Transactions on Engineering Management*, Vol. 46, No. 2, 1999, pp. 132–143.
162. Kawagoe, M., Yamaguchi, T. and Aoyama, H., TOPICA Project: Towards the Total System for Presentation and Invention by Creativity-Acceleration, *Proceedings of the Hawaii International Conference on System Science*, Vol. 4, 1992, pp. 389–397.
163. Keeney, R.L., Creativity in Decision-Making with Value-Focused Thinking, *Sloan Management Review*, Vol. 35, No. 4, 1994, pp. 33–41.
164. Kickul, J. and Gundry, L.K., Breaking Through Boundaries for Organizational Innovation: New Managerial Roles and Practices in e-Commerce Firms, *Journal of Management*, Vol. 27, No. 3, 2001, pp. 347–361.

165. Kim, J., *Creativity*, Scholastic, New York, 2002.
166. Kimball, R.S., *The Winds of Creativity: Finding Fulfillment Through Creative Act*, Green Timber Publications, Portland, Maine, 1996.
167. Kirst, W. and Diekmeyer, U., *Creativity Training: Become Creative in 30 Minutes a Day*, P.H. Wyden, New York, 1973.
168. Kivett, H.A., Fusion of Creativity in Rail Transit Stations: A Retrospective and Critique, *Transportation Research Record*, No. 1549, 1996, pp. 75–78.
169. Kjellberg, A. and Werneman, A., Business Innovation — Innovative Teams, Competence Brokers and Beehive Structures — in a Sustainable Work Organization, *Cirp Annals*, Vol. 49, No. 1, 2000, pp. 355–358.
170. Klein, E.E. and Dologite, D.G., Role of Computer Support Tools and Gender Composition in Innovative Information System Idea Generation by Small Groups, *Computers in Human Behavior*, Vol. 16, No. 2, 2000, pp. 111–139.
171. Kletke, M.G., Mackay, J.M., Barr, S.H. and Johnes, B., Creativity in the Organization: The Role of Individual Creative Problem Solving and Computer Support, *International Journal of Human-Computer Studies*, Vol. 55, No. 3, 2001, pp. 217–237.
172. Kor, Y.Y. and Mahoney, J.T., Penrose's Resource-Based Approach: The Process and Product of Research Creativity, *Journal of Management Studies*, Vol. 37, No. 1, 2000, pp. 109–139.
173. Krettek, T. (editor), *Creativity and Common Sense: Essays in Honor of Paul Weiss*, State University of New York Press, Albany, New York, 1987.
174. Kryssanov, V.V., Tamaki, H. and Kitamura, S., Towards a Theory of Creativity for Engineering Design, *Proceedings of the IEEE International Conference on Systems, Man and Cybernetics*, Vol. 1, 1999, pp. I-250–I-255.
175. Kryssanov, V.V., Tamaki, H. and Kitamura, S., Understanding Design Fundamentals: How Synthesis and Analysis Drive Creativity, Resulting in Emergence, *Artificial Intelligence in Engineering*, Vol. 15, No. 4, 2001, pp. 329–342.
176. Kuecken, J.A., *Creativity, Invention, and Progress*, H.W. Sams, Indianapolis, Indiana, 1969.
177. Kurtzberg, T.R., Creative Thinking, Cognitive Aptitude, and Integrative Joint Gain: A Study of Negotiator Creativity, *Creativity Research Journal*, Vol. 11, No. 4, 1998, pp. 283–293.
178. Kuryllowicz, K., Creativity and Innovation Crucial to Selling in Today's Marketplace, *Pulp and Paper Journal*, Vol. 43, No. 10, 1990, p. 4.
179. Lakin, P., *Creativity: Around the World*, Blackbirch Press, Woodbridge, Connecticut, 1995.
180. Laplante, P. and Flaxman, H., Convergence of Technology and Creativity in the Corporate Environment, *IEEE Transactions on Professional Communication*, Vol. 38, No. 1, 1995, pp. 20–23.
181. Larson, M.C., Thomas, B.H. and Leviness, P.O., Assessing Creativity in Engineers, *American Society of Mechanical Engineers, Design Engineering Division (Publication)*, Vol. 102, 1999, pp. 1–6.

182. Lee, H. and Choi, B., Knowledge Management Enablers, Processes, and Organizational Performance: An Integrative View and Empirical Examination, *Journal of Management Information Systems*, Vol. 20, No. 1, 2003, pp. 179–228.
183. Leech, T.G., Quantifying the Creative Process, *Concrete International*, Vol. 17, No. 8, 1995, pp. 54–57.
184. Leenders, R.T.A.J., Van Engelen, J.M.L. and Kratzer, J., Virtuality, Communication, and New Product Team Creativity: A Social Network Perspective, *Journal of Engineering and Technology Management — Jet-M*, Vol. 20, Nos. 1–2, 2003, pp. 69–92.
185. Lerner, H., *Creativity*, Harper Collins Publishers, New York, 1993.
186. Levine, R., Creativity Through Applied Science, *Modern Steel Construction*, Vol. 33, No. 6, 1993, pp. 14–19.
187. Li, L. *et al.*, Measurement of Management Creativity, *International Journal of Psychology*, Vol. 35, Nos. 3–4, 2000, p. 449.
188. Liu, Y.T., Personal versus Cultural Cognitive Models of Design Creativity, *International Journal of Technology and Design Education*, Vol. 8, No. 2, 1998, pp. 185–195.
189. Lobert, B.M. and Dologite, D.G., Measuring Creativity of Information System Ideas: An Exploratory Investigation, *Proceedings of the Hawaii International Conference on System Sciences*, Vol. 4, 1994, pp. 392–402.
190. Low, M.K., Lamvik, T., Walsh, K. and Myklebust, O., Product to Service Eco-Innovation: The TRIZ Model of Creativity Explored, *Proceedings of the IEEE International Symposium on Electronics and the Environment*, 2000, pp. 209–214.
191. Lowe, P., *Creativity and Problem Solving*, McGraw-Hill, New York, 1995.
192. Maccrimmon, K.R. and Wagner, C., Stimulating Ideas Through Creativity Software, *Management Science*, Vol. 40, No. 11, 1994, pp. 1514–1532.
193. Maher, M., Lou, Z., Fang, G. and John S., Creativity in Humans and Computers, *Acta Polytechnica Scandinavica, Civil Engineering and Building Construction Series*, No. 92, 1989, pp. 129–141.
194. Maimon, O.Z. and Horowitz, R., Sufficient Conditions for Inventive Solutions, *IEEE Transactions on Systems, Man and Cybernetics, Part C: Applications and Reviews*, Vol. 29, No. 3, 1999, pp. 349–361.
195. Marakas, G.M. and Elam, J.J., Creativity Enhancement in Problem Solving: Through Software or Process?, *Management Science*, Vol. 43, No. 8, 1997, pp. 1136–1146.
196. Mascitelli, R., From Experience: Harnessing Tacit Knowledge to Achieve Breakthrough Innovation, *Journal of Product Innovation Management*, Vol. 17, No. 3, 2000, pp. 179–193.
197. Mauzy, J. and Harriman, R.A., Three Climates for Creativity, *Research-Technology Management*, Vol. 46, No. 3, 2003, pp. 27–30.
198. Mauzy, J. and Harriman, R., *Building an Inventive Organization*, Harvard Business School Press, Boston, Massachusetts, 2003.

199. May, D.R. and Riolli, L., Aging and the Creative Process: The Influence of Aging, Age Similarity, and Age Stereotypes on the Generation and Adoption of Creative Ideas, *Proceedings of the Annual Meeting of the Decision Sciences Institute*, Vol. 1, 1996, p. 339.
200. McAdam, R. and McClelland, J., Sources of New Product Ideas and Creativity Practices in the UK Textile Industry, *Technovation*, Vol. 22, No. 2, 2002, pp. 113–121.
201. McIntyre, S.C., Higgins, L.F. and Couger, J.D., Knowledge Base Enrichment via Object Oriented Creativity Techniques, *Proceedings of the Hawaii International Conference on System Science*, Vol. 3, 1992, pp. 373–381.
202. McLeod, P.L., Lobel, S.A. and Cox, T.H., Ethnic Diversity and Creativity in Small Groups, *Small Group Research*, Vol. 27, No. 2, 1996, pp. 248–264.
203. McPherson, J.H., *How are they, are we, am I, Doing?*, Pendell Pub. Co., Midland, Michigan, 1968.
204. Micklus, C.S., *Creativity + Teamwork = Solutions*: *Creative Competitions*, Glassboro, New Jersey, 1997.
205. Middelmann, U., Innovation as a Strategic Success Factor, *Technische Mitteilungen Krupp* (*English Edition*), No. 1, 1995, pp. 7–12.
206. Milgram, R.M. (editor), *Counseling Gifted and Talented Children*: *A Guide for Teachers, Counselors, and Parents*, Ablex Publ. Corp., Norwood, New Jersey, 1991.
207. Mohamed, M.A.K., Assessing Determinants of Departmental Innovation — An Exploratory Multi-Level Approach, *Personnel Review*, Vol. 31, Nos. 5–6, 2002, pp. 620–641.
208. Morash, C. (editor), *Creativity and Its Contexts*, Lilliput Press, Dublin, 1995.
209. Morgan, T.F. and Ammentorp, W.M., Practical Creativity in the Corporate World — Capturing Expert Judgment with Qualitative Models, *American Behavioral Scientist*, Vol. 37, No. 1, 1993, pp. 102–111.
210. Mueller, S.L. and Thomas, A.S., Culture and Entrepreneurial Potential: A Nine Country Study of Locus of Control and Innovativeness, *Journal of Business Venturing*, Vol. 16, No. 1, 2001, pp. 51–75.
211. Mumford, M.D. and Moertl, P., Cases of Social Innovation: Lessons from Two Innovations in the 20th Century, *Creativity Research Journal*, Vol. 15, Nos. 2–3, 2003, pp. 261–266.
212. Mumford, M.D., Social Innovation: Ten Cases from Benjamin Franklin, *Creativity Research Journal*, Vol. 14, No. 2, 2002, pp. 253–266.
213. Murray, C.J. and Gottschalk, M.A., Services. Where Creativity Thrives, *Design News* (*Boston*), Vol. 49, No. 13, 1993, p. 7.
214. Nagasundaram, M. and Bostrom, R.P., Structuring of Creative Processes: Implications for GSS Research, *Proceedings of the Hawaii International Conference on System Sciences*, Vol. 4, 1994, pp. 51–60.
215. Nagel, S. (editor), *Creativity*: *Being Usefully Innovative in Solving Diverse Problems*, Nova Science Publishers, Huntington, New York, 2000.

216. Nakakoji, K., Yamamoto, Y. and Ohira, M., Computational Support for Collective Creativity, *Knowledge-Based Systems*, Vol. 13, Nos. 7–8, 2000, pp. 451–458.
217. Nambisan, S., Agarwal, R. and Tanniru, M., Organizational Mechanisms for Enhancing User Innovation in Information Technology, *MIS Quarterly*, Vol. 23, No. 3, 1999, pp. 365–395.
218. Navin, F.P.D., Engineering Creativity — Doctum Ingenium, *Canadian Journal of Civil Engineering*, Vol. 21, No. 3, 1994, pp. 499–511.
219. Nemeth, C.J., Managing Innovation: When Less is More, *IEEE Engineering Management Review*, Vol. 26, No. 1, 1998, pp. 58–66.
220. Nickel, T.M. and Krems, J.F., Leadership and Creativity: An Empirical Study on Factors Affecting Work-Related Suggestions Made by Employees, *Zeitschrift Fur Arbeits-Und Organisationspsychologie*, Vol. 42, No. 1, 1998, pp. 27–32.
221. Nijhof, A., Krabbendam, K. and Looise, J.C., Innovation Through Exemptions: Building Upon the Existing Creativity of Employees, *Technovation*, Vol. 22, No. 11, 2002, pp. 675–683.
222. Novoa-Weber, C., Carle, D., Liening, A. and Keys, E., Product Innovation Process, *IEE Colloquium (Digest)*, No. 442, 1998, p. 9/2.
223. Nunamaker, J.F., Applegate, L.M. and Konsynski, B.R., Facilitating Group Creativity: Experience with a Group Decision Support System, *Proceedings of the 20th Hawaii International Conference on System Sciences*, Vol. 1, 1995, pp. 422–430.
224. Nystrom, H., *Creativity and Innovation*, Wiley, New York, 1979.
225. O'Connor, G.C. and Rice, M.P., Opportunity Recognition and Breakthrough Innovation in Large Established Firms, *California Management Review*, Vol. 43, No. 2, 2001, pp. 95–102.
226. Ogilvie, D.T., Creative Action as a Dynamic Strategy: Using Imagination to Improve Strategic Solutions in Unstable Environments, *Journal of Business Research*, Vol. 41, No. 1, 1998, pp. 49–56.
227. Osborn, S. and Elliott, G., Standard Creativity: Creating Flexible Web Development Standards, *Proceedings of the IEEE International Professional Communication Conference*, 2002, pp. 1–21.
228. Padival, N., Techniques for Administrative Creativity in Local Government, *Journal of Management in Engineering*, Vol. 14, No. 1, 1998, pp. 20–21.
229. Paper, D., Value of Creativity in Business Process Reengineering, *Proceedings of the Hawaii International Conference on System Sciences*, Vol. 3, 1997, pp. 290–297.
230. Parden, R.J., Focusing the Knowledge Management Process, *PICMET*, 2001, pp. 253–256.
231. Payne, R.L., Creativity and the Management of Change, *R & D Management*, Vol. 30, No. 4, 2000, pp. 374–376.
232. Perry, T.S., Designing a Culture for Creativity, *Research-Technology Management*, Vol. 38, No. 2, 1995, pp. 14–17.
233. Petrucelli, R., *Learn the Value of Creativity*, Rourke Enterprises, Vero Beach, Florida, 1989.

234. Plsek, P.E., *Creativity, Innovation, and Quality*, ASQ Quality Press, Milwaukee, Wisconsin, 1997.
235. Plsek, P.E., Incorporating the Tools of Creativity into Quality Management, *Quality Progress*, Vol. 31, No. 3, 1998, pp. 21–28.
236. Plsek, P., Bringing Creativity to the Pursuit of Quality, *Annual Quality Congress Transactions*, 1996, pp. 99–105.
237. Plsek, P., Creativity in Process Redesign and Reengineering: An Interactive Tutorial, *Annual Quality Congress Transactions*, 1997, pp. 954–956.
238. Plsek, P.E., Incorporating the Tools of Creativity into Quality Management, *Quality Progress*, Vol. 31, No. 3, 1998, pp. 21–28.
239. Plsek, Paul E., Incorporating the Tools of Creativity into Quality Management, *IEEE Engineering Management Review*, Vol. 26, No. 3, 1998, pp. 61–68.
240. Porter, R.J., Pleau, R. and Hall, B.R., Software "Workbench" Promotes Creativity, *Research & Development*, Vol. 29, No. 8, 1987, pp. 72–75.
241. Provost, L.P. and Langley, G.J., Importance of Concepts in Creativity and Improvement, *Quality Progress*, Vol. 31, No. 3, 1998, pp. 31–38.
242. Provost, L.P. and Sproul, R.M., Creativity and Improvement: A Vital Link, *Quality Progress*, Vol. 29, No. 8, 1996, pp. 101–107.
243. Raine, J.K., Design Innovation and Project Engineering: Paths to Profit, *Transactions of the Institution of Professional Engineers New Zealand, Electrical/Mechanical/Chemical Engineering Section*, Vol. 13, No. 2, 1986, pp. 95–105.
244. Ralston, D.W. and Nadler, G., Breakthrough Thinking. A New Problem Solving Paradigm for Total Quality Management, *Proceedings of the International Industrial Engineering Conference*, 1998, pp. 151–158.
245. Ramirez, M.R., Meaningful Theory of Creativity: Design as Knowledge: Implications for Engineering Design, *Proceedings of the Frontiers in Education Conference*, 1994, pp. 594–597.
246. Ramirez, M.R., Engineering Vision: Considerations in a Meaningful Approach to Conceptual Design, *AI Edam-Artificial Intelligence for Engineering Design Analysis and Manufacturing*, Vol. 10, No. 3, 1996, pp. 199–214.
247. Rasch, M., Computer Based Instructional Strategies to Improve Creativity, *Computers in Human Behavior*, Vol. 4, No. 1, 1988, pp. 23–28.
248. Rasmus, D.W., Creativity and Tools, *PC AI Intelligent Solutions for Desktop Computers*, Vol. 9, No. 3, 1995, pp. 29–31.
249. Ray, M.L. and Myers, R., *Creativity in Business*, Doubleday, New York, 1989.
250. Razik, T.A., *Bibliography of Creativity Studies and Related Areas*, State University of New York at Buffalo Press, Buffalo, 1965.
251. Richards, L.G., Stimulating Creativity: Teaching Engineers to be Innovators, *Proceedings of the Frontiers in Education Conference*, Vol. 3, 1998, pp. 1034–1039.
252. Rickards, T., *Creativity and the Management of Change*, Blackwell Business, Malden, Massachusetts, 1999.

253. Rickards, T., *Creativity at Work*, Aldershot, Hantshire, England, 1988.
254. Riley, H.W., Promoting Creativity in the Electric Utility Industry Under a Regulated and/or De-Regulated Environment, *American Society of Mechanical Engineers, Environmental Control Division Publication, EC*, Vol. 1, 1996, pp. 227–232.
255. Ristola, K., Creative Problem Solving in Architectural Design, *Acta Polytechnica Scandinavica, Civil Engineering and Building Construction Series*, No. 92, 1989, pp. 119–128.
256. Robertson, S., Requirements Trawling: Techniques for Discovering Requirements, *International Journal of Human-Computer Studies*, Vol. 55, No. 4, 2001, pp. 405–421.
257. Rosen, E.A. and Ruzicka, M.F., What Makes an Engineer Creative?, *Proceedings of the International Engineering Management Conference*, 1990, pp. 281–284.
258. Rouibah, K. and Ould-ali, S., PUZZLE: A Concept and Prototype for Linking Business Intelligence to Business Strategy, *Journal of Strategic Information Systems*, Vol. 11, No. 2, 2002, pp. 133–152.
259. Rouse, W.B., A Note on the Nature of Creativity in Engineering — Implications for Supporting System-Design, *Information Processing and Management*, Vol. 22, No. 4, 1986, pp. 279–285.
260. Roweton, W.E., Creativity: A Review of Theory and Research, Wisconsin Research and Development Center for Cognitive Learning, Madison, Wisconsin, 1970.
261. Runco, M.A., *Divergent Thinking*, Ablex Publ. Corp., Norwood, New Jersey, 1991.
262. Runco, M.A. (editor), *Problem Finding, Problem Solving, and Creativity*, Ablex Publ. Corp., Norwood, New Jersey, 1994.
263. Runco, M.A., *The Creativity Research Handbook*, Hampton Press, Cresskill, New Jersey, 1997.
264. Russell, E., Creating the Right Climate, *New Electronics*, Vol. 34, No. 19, 2001, pp. 23–24.
265. Saaty, T.L., Reflections and Projections on Creativity in Operations Research and Management Science: A Pressing Need for a Shift in Paradigm, *Operations Research*, Vol. 46, No. 1, 1998, pp. 9–16.
266. Santanen, E.L., Briggs, R.O. and De Vreede, G.J., Cognitive Network Model of Creativity: A New Causal Model of Creativity and a New Brainstorming Technique, *Proceedings of the Hawaii International Conference on System Sciences*, 2000, p. 171.
267. Sawyer, R.K. (editor), *Creativity in Performance*, Ablex Publishing Corporation, Greenwich, Connecticut, 1997.
268. Sawyer, R.K. *et al.*, *Creativity and Development*, Oxford University Press, New York , 2003.
269. Schank, R.C. and Foster, D.A., Engineering of Creativity: A Review of Boden's the Creative Mind, *Artificial Intelligence*, Vol. 79, No. 1, 1995, pp. 129–143.

270. Schwarz, K.K., Methodical Creativity, *Manufacturing Engineer*, Vol. 77, No. 1, 1998, pp. 17–20.
271. Scott, R.K., Creative Employees — A Challenge to Managers, *Journal of Creative Behavior*, Vol. 29, No. 1, 1995, pp. 64–71.
272. Scratchley, L.S. and Hakstian, A.R., The Measurement and Prediction of Managerial Creativity, *Creativity Research Journal*, Vol. 13, Nos. 3–4, 2000, pp. 367–384.
273. Scriabin, M., Kotak, D.B. and Whale, K.G., Symbiotic Systems: Exploiting Human Creativity, *European Journal of Operational Research*, Vol. 84, No. 2, 1995, pp. 227–234.
274. Sebba, G., *Creativity: Lectures*, Scholars Press, Atlanta, Georgia, 1987.
275. Seijts, G.H. and Latham, B.W., Creativity Through Applying Ideas from Fields Other than One's Own: Transferring Knowledge from Social Psychology to Industrial/Organizational Psychology, *Canadian Psychology-Psychologie Canadienne*, Vol. 44, No. 3, 2003, pp. 232–239.
276. Sethi, R., Smith, D.C. and Park, C.W., Cross-Functional Product Development Teams, Creativity, and the Innovativeness of New Consumer Products, *Journal of Marketing Research*, Vol. 38, No. 1, 2001, pp. 73–85.
277. Shane, S.A., Why do Some Societies Invent More than Others, *Journal of Business Venturing*, Vol. 7, No. 1, 1992, pp. 29–46.
278. Shapiro, G., Employee Involvement: Opening the Diversity Pandora's Box?, *Personnel Review*, Vol. 29, No. 3, 2000, pp. 304–323.
279. Shaw, M.P., On the Creative Process in Science and Engineering, *Proceedings of the Portland Int. Conf. on Manage. Eng. Technol.*, 1991, pp. 635–639.
280. Shaw, M.P. and Runco, M.A. (editors), *Creativity and Affect*, Ablex Publ. Corp., Norwood, New Jersey, 1994.
281. Shaw, T., Arnason, K. and Belardo, S., Effects of Computer Mediated Interactivity on Idea Generation: An Experimental Investigation, *IEEE Transactions on Systems, Man and Cybernetics*, Vol. 23, No. 3, 1993, pp. 737–745.
282. Sheasley, W.D., Leading the Technology Development Process, *Research-Technology Management*, Vol. 42, No. 3, 1999, pp. 49–55.
283. Shlaes, C., Rewarding and Stimulating Creativity and Innovation in Technology Companies, *Proceedings of the Portland Int. Conf. on Manage. Eng. Technol.*, 1991, pp. 609–612.
284. Shneiderman, B., Supporting Creativity with Powerful Composition Tools for Artifacts and Performances, *Proceedings of the Hawaii International Conference on System Sciences*, 2000, p. 172.
285. Simberg, A.L., Creativity at Work: The Practical Application of a Complete Program, Industrial Education Institute, Boston, 1964.
286. Simonton, D.K., *Creativity in Science: Change, Logic, Genius, and Zeitgeist*, Cambridge University Press, New York, 2004.
287. Steptoe, J., *Creativity*, Clarion Books, New York, 1997.
288. Steuver, J.K., Creativity: A Response to Change and Competition, *SAVE Proceedings (Society of American Value Engineers)*, Vol. 27, 1999, pp. 209–214.

289. Stevens, G., Burley, J. and Divine, R., Creativity Plus Business Discipline Equals Higher Profits Faster from New Product Development, *Journal of Product Innovation Management*, Vol. 16, No. 5, 1999, pp. 455–468.
290. Streitz, N.A., Geissler, J., Holmer, T., Konomi, S., Mueller-Tomfelde, C., Reischl, W., Rexroth, P., Seitz, P. and Steinmetz, R., LAND: An Interactive Landscape for Creativity and Innovation, *Proceedings of the Conference on Human Factors in Computing Systems*, 2000, pp. 120–127.
291. Strzalecki, A., Creativity in Design — General Model and its Verification, *Technological Forecasting and Social Change*, Vol. 64, Nos. 2–3, 2000, pp. 241–260.
292. Subotnik R. *et al.* (editors), *Genius Revisited: High IQ Children Grown up*, Ablex Publ. Corp., Norwood, New Jersey, 1993.
293. Subotnik, R.F. and Arnold, K.D. (editors), *Beyond Terman: Contemporary Longitudinal Studies of Giftedness and Talent*, Ablex Pub. Corp., Norwood, New Jersey, 1994.
294. Sugiomoto, M., Hori, K. and Ohsuga, S., Method of Assisting Creative Design Processes, *Languages of Design*, Vol. 1, No. 4, 1993, pp. 357–367.
295. Sutton, RI., The Weird Rules of Creativity, *Harvard Business Review*, Vol. 79, No. 8, 2001, pp. 94–99.
296. Sweetman, K., Management Mistakes Squelch Employee Innovation — A New Experiment Reveals that Even the Best Managers Often Suppress Workplace Creativity — Sometimes by Trying to be Helpful, *MIT Sloan Management Review*, Vol. 42, No. 4, 2001, pp. 9–10.
297. Takesue, T., CG Technologies for Supporting Cooperative Creativity by Industrial Designers, *Proceedings of the IEEE International Workshop on Robot and Human Communication*, 2000, pp. 316–321.
298. Tang, H.K., Integrative Model of Innovation in Organizations, *Technovation*, Vol. 18, No. 5, 1998, pp. 297–309.
299. Taylor, C.W., *Creativity: Progress and Potential*, McGraw-Hill, New York, 1964.
300. Tennyson, R.D. and Breuer, K., Improving Problem Solving and Creativity Through Use of Complex-Dynamic Simulations, *Computers in Human Behavior*, Vol. 18, No. 6, 2002, pp. 650–668.
301. Tesluk, P.E., Farr, J.L. and Klein, S.R., Influences of Organizational Culture and Climate on Individual Creativity, *Journal of Creative Behavior*, Vol. 31, No. 1, 1997, pp. 27–41.
302. Thamhain, H.J., Can Innovative R & D Performance be Managed, *PICMET*, 2001, pp. 151–159.
303. Thamhain, H.J., Managing Innovative R & D Teams, *R & D Management*, Vol. 33, No. 3, 2003, pp. 297–311.
304. Thomas, C.A., *Creativity in Science*, Massachusetts Institute of Technology Press, Boston, 1955.
305. Thompson, G. and Lordan, M., Review of Creativity Principles Applied to Engineering Design, *Proceedings of the Institution of Mechanical Engineers. Part E, Journal of Process Mechanical Engineering*, Vol. 213, No. 1, 1999, pp. 17–31.

306. Tomas, S., Creative Problem-Solving: An Approach to Generating Ideas, *Proceedings of the Annual American Production and Inventory Control Society International Conference*, 1998, pp. 450–452.
307. Tomas, Sam, Creative Problem-Solving: An Approach to Generating Ideas, *Proceedings of the Annual American Production and Inventory Control Society International Conference*, 1997, pp. 450–455.
308. Torlak, G.N., Rationalization of Metaphorical Exploration: Improving the Creativity Phase of Total Systems Intervention (TSI) on the Basis of Theory and Practice, *Systemic Practice and Action Research*, Vol. 14, No. 4, 2001, pp. 451–482.
309. Torrance, E.P., *Why Fly?*, Ablex Publ. Corp., Norwood, New Jersey, 1995.
310. Van Der Lugt, R., Developing a Graphic Tool for Creative Problem Solving in Design Groups, *Design Studies*, Vol. 21, No. 5, 2000, pp. 505–522.
311. Van Eijnatten, F.M. and Simonse, L.W.L., Organizing for Creativity, Quality and Speed in Product Creation Processes, *Quality and Reliability Engineering International*, Vol. 15, No. 6, 1999, pp. 411–416.
312. Vansuch, G.M., CIA for CS: Creativity and Innovation Applications for Customer/Supplier Relationships, *Annual Quality Congress Transactions*, 1998, pp. 447–451.
313. Vaucelle, C. and Jehan, T., Dolltalk: A Computational Toy to Enhance Children's Creativity, *Proceedings of the Conference on Human Factors in Computing Systems*, 2002, pp. 776–777.
314. Vernon, P.E. (editor), *Creativity: Selected Readings*, Penguin, New York, 1970.
315. Verstijnen, I.M., Stuyver, R., Hennessey, J.M., Van Leeuwen, C.C. and Hamel, R., Considerations for Electronic Idea-Creation Tools, *Proceedings of the Conference on Human Factors in Computing Systems*, 1996, pp. 197–198.
316. Vzyatishev, V.F., Engineering Design and Creativity, *International Journal of Continuing Engineering Education*, Vol. 1, No. 3, 1991, pp. 219–234.
317. Wakefield, J.F., *Creative Thinking: Problem Solving Skills and the Arts Orientation*, Ablex Pub. Corp., Norwood, New Jersey, 1992.
318. Walker, S., *Queensberry Hunt: Creativity and Industry*, Fourth Estate and Wordsearch, London, 1992.
319. Wang, C.W., Wu, J.J. and Horng, R.Y., Creative Thinking Ability, Cognitive Type and R & D Performance, *R & D Management*, Vol. 29, No. 3, 1999, pp. 247–254.
320. Ward, T.B., Finke, R.A. and Smith, S.M., *Creativity and the Mind: Discovering the Genius Within*, Plenum Press, New York, 1995.
321. Weber, R.J., Toward a Language of Invention and Synthetic Thinking, *Creativity Research Journal*, Vol. 9, No. 4, 1996, pp. 353–367.
322. Weisberg, R.W., *Creativity: Beyond the Myth of Genius*, W.H. Freeman, New York, 1993.
323. Weisberg, R.W., *Creativity: Genius and Other Myths*, W.H. Freeman, New York, 1986.

324. Wen, G.H., Zheng, Q.L. and Pan, D., Theoretical Analysis of Creative Methods, *Proceedings of the IEEE International Conference on Systems, Man and Cybernetics*, Vol. 5, 2001, pp. 2811–2816.
325. West, M.A. and Anderson, N.R., Innovation in Top Management Teams, *Journal of Applied Psychology*, Vol. 81, No. 6, 1996, pp. 680–693.
326. West, M.A., Sparkling Fountains or Stagnant Ponds: An Integrative Model of Creativity and Innovation Implementation in Work Groups, *Applied Psychology — An International Review*, Vol. 51, No. 3, 2002, pp. 355–387.
327. White, D., Stimulating Innovative Thinking, *Research-Technology Management*, Vol. 39, No. 5, 1996, pp. 31–35.
328. Whitfield, P.R., *Creativity in Industry*, Penguin, London, 1975.
329. Williams, A., *Creativity, Invention and Innovation: A Guide to Building your Business Future*, Allen & Unwin, St. Leonards, New South Wales, Australia, 1999.
330. Williams, S.D., Self-Esteem and the Self-Censorship of Creative Ideas, *Personnel Review*, Vol. 31, No. 4, 2002, pp. 495–503.
331. Williamson, B., Creativity, the Corporate Curriculum and the Future: A Case Study, *Futures*, Vol. 33, No. 6, 2001, pp. 541–555.
332. Wishart, T., *Sun: Creativity and Environment*, Universal Edition, London, 1974.
333. Wong, S. and Pang, L., Motivators to Creativity in the Hotel Industry — Perspectives of Managers and Supervisors, *Tourism Management*, Vol. 24, No. 5, 2003, pp. 551–559.
334. Wood, M., Creativity and the Management of Change, *Organization Studies*, Vol. 21, No. 5, 2000, pp. 1013–1015.
335. Woods, M.S., *Creativity: Process and Product*, Seven C's, Seattle, 1977.
336. Wright, S., Innovation and Creativity, *Paper Age*, Vol. 11, No. 6, 1999, p. 16.
337. Yamaguchi, T., Sato, T., Wakamatsu, Y. and Hashimoto, T., Creativity Support Using Chaotic Retrieval on Fuzzy Associative Memory System, *Proceedings of the IEEE International Conference on Systems, Man and Cybernetics*, Vol. 3, 1996, pp. 1966–1971.
338. Yanagishita, K. and Ishii, K., Training of New Product Development for Industrial Engineering Students: A Case from the Kanazawa Institute of Technology, *International Journal of Technology Management*, Vol. 25, Nos. 6–7, 2003, pp. 659–665.
339. Yeo, R., An Integrative Approach to the Teaching of Technical Communication Skills, *Innovations in Education and Training International*, Vol. 38, No. 1, 2001, pp. 93–100.
340. Yep, D.S.M., *Creativity at Work,* Irwin Professional Pub., Burr Ridge, Illinois, 1994.
341. Yong, L.M.S., Managing Creative People, *Journal of Creative Behavior*, Vol. 28, No. 1, 1994, pp. 16–20.

# Index